C3
建筑立场系列丛书 No.19
U0332341
建筑入景
Add in the Scape
中文版
韩国C3出版公社 | 编
赵姗姗 王思锐 时跃 陈帅甫 葛永宏 胡筱狄 牛文佳 张琳娜 | 译
大连理工大学出版社

03

建筑立场系列丛书 No.19

4

资讯

20

建筑入景

96

嵌入场地的建筑

154

EXIT建筑师事务所

News

20

Add in the Scape

96

Nestle in

154

EXIT ARCHITECTS

锡拉库扎古希腊剧院

西西里岛锡拉库扎希腊剧院的舞台设计是由OMA完成的，于2012年5月11日正式开幕，首演剧目为埃斯库罗斯的《解放了的普罗米修斯》（由克劳迪奥·隆吉导演）。舞台设有三台临时建筑装置，对这个始建于公元前5世纪的剧院空间进行了重新诠释。OMA的干预措施将在今年夏天的表演周期内得到最大化的开发和适应，这个周期的表演由Istituto Nazionale del Dramma Antico组织，剧目包括欧里庇得斯的《酒神的女伴》（由Antonio Calenda导演）和阿里斯托芬的《鸟》（由罗伯塔·托雷导演）。OMA的第一个干预措施是"环"，这是一条悬浮的走道，与梯级座位形成半圆形，围绕着舞台和后台，为演员增加了一种入场方式。"机器"对戏剧表演而言是一个完全合适的背景：一个7m高的倾斜圆形平台与圆形露天剧场形成镜像关系。该背景可以旋转，象征着普罗米修斯承受折磨期间13个世纪的通道。它的中间是分开的，也可以被打开，作为演员的入口处，并象征着戏剧性事件，如普罗米修斯被吞噬进地球深处。"筏"是为演员和舞者设计的圆形舞台，它使管弦乐队空间重新成为现代thymele——古代专门用于酒神仪式的祭坛空间。OMA设计创新型表演空间已经有很长的历史，希腊剧院舞台设计只是其中的一部分，他们的设计包括荷兰舞蹈剧院（1987年）和达拉斯的威利剧院（与Rex共同设计，2009年），以及台北表演艺术中心——三个适应性极强的剧院被融合到一个中央立方体中，现正在中国台湾进行建设。

Ancient Greek Theatre in Syracuse_OMA

OMA's design for the stage set at the Greek Theatre in Syracuse, Sicily, was inaugurated on 11 May 2012 with the performance of *Aeschylus's Prometheus Unbound* (directed by Claudio Longhi). The scenography features three temporary architectural devices that reinterpret the spaces of the theatre, which dates from the 5th century BCE. OMA's interventions will be dramatically exploited and adapted at strategic moments within this summer's cycle of plays staged by the Istituto Nazionale del Dramma Antico, which also includes *Euripides' Bacchae* (dir. Antonio Calenda) and Aristophane's *The Birds* (dir. Roberta Torre). The first intervention, the Ring, is a suspended walkway that completes the semi-circle of the terraced seating, encompassing the stage and the backstage, and giving actors an alternative way of entering the scene. The Machine is a fully adaptable backdrop for the plays: a sloping circular platform, seven metres high, mirroring the amphitheatre. The backdrop can rotate, symbolizing the passage of 13 centuries during Prometheus's torture; split down the middle, it can also be opened, allowing the entrance of the actors, and symbolizing dramatic events like the Prometheus being swallowed in the bowels of the earth. The Raft, a circular stage for the actors and dancers, reimagines the orchestra space as a modern thymele, an

©MeRyan

altar that in ancient times was dedicated to Dionysian rites. The Greek Theatre scenography is part of the office's long history of designing innovative performance spaces, from the Netherlands Dance Theatre (1987) and the Wyly Theatre in Dallas (with Rex, 2009), to the Taipei Performing Arts Centre – three adaptable theatres plugged into a central cube, now under construction in Taiwan, China.

项目名称：Stage Set for Ancient Greek Theatre in Syracuse
地点：Teatro Greco, Syracuse, Italy
主管合伙人：Rem Koolhaas
副主管：Ippolito Pestellini Laparelli
项目建筑师：Francesco Moncada
项目团队：Miguel Taborda, Barbara Materia
当地建筑师：Emanuela Reale (I.N.D.A.)
结构工程师：Sebastiano Floridia
脚手架与金属结构工程师：Mauro Ianno', Renato Ianno' C.S.M.I. Srl Melilli
木覆层：Laboratorio Scenico I.N.D.A.
甲方：Istituto Nazionale del Dramma Antico (INDA)
用地面积：8,374m² 建筑面积：2,358m²
总楼面面积：2,810m² 竣工时间：2012
摄影师：©Alberto Moncada (courtesy of the architect) (except as noted)

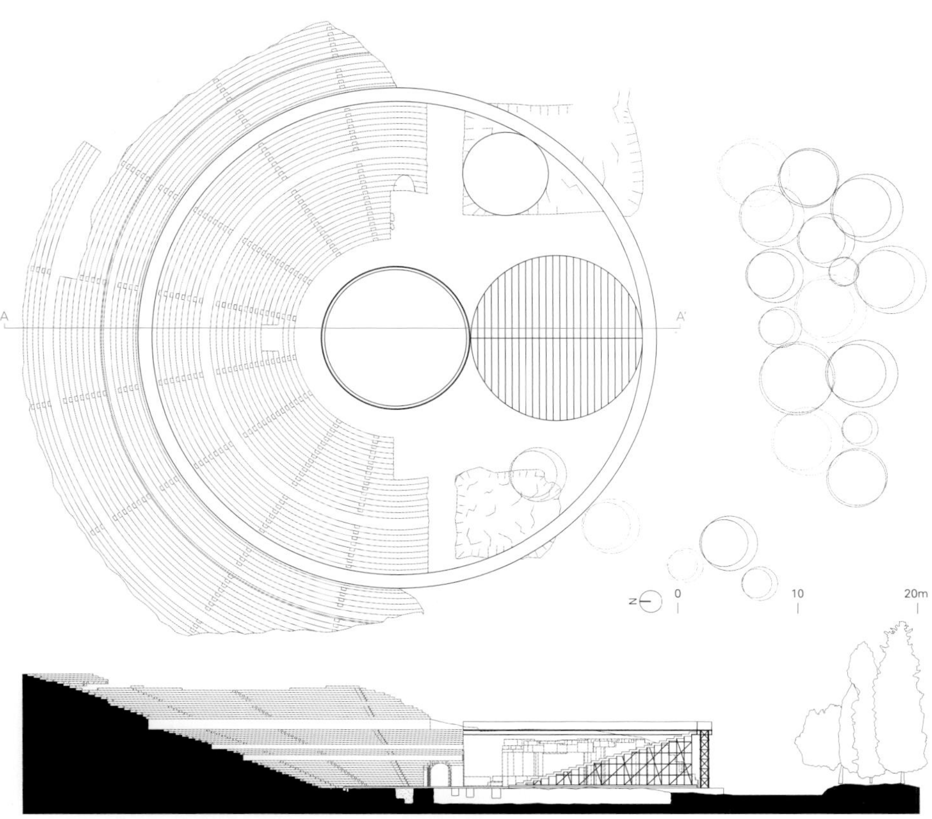

A-A' 剖面图 section A-A'

视角寺庙

写作《无神论者的宗教》时，阿兰·德波顿提出了各种切实可行的方法，从中我们也许能够从宗教的角度出发改变世俗的现代世界。

他提出，无神论者应该学习复制主要宗教，建造寺庙。他还建议无神论者组织起来，在各城市中建立一个新的寺庙网络。

该提案是德波顿众多争论中的一部分，德波顿认为无神论者应停止对宗教的消极态度，体验人们从宗教中获得的快乐。然后，他们应该复制它，这完全与上帝无关。

德波顿认为，像理查德·道金斯这样的无神论者永远都不会让人相信无神论是一种看待生活的有吸引力的方式，除非他们能提供宗教一直使用的某种仪式、建筑、社区以及艺术作品。

德波顿已经采取了第一步，使寺庙项目成为现实。他已经与Tom Greenall建筑师事务所以及Jordan Hodgson合作设计出第一座寺庙，以后还将有不同的建筑师设计的众多寺庙，共同形成理想的寺庙链。

这第一个结构被称为视角寺庙，设计旨在让人们恢复生活中、在多种刺激和现代世界混乱之中的角度感。

视角寺庙将在伦敦金融城建造，它将是一个黑色塔形结构，其结构代表了地球的年龄，每厘米相当于100万年。该建筑物总高度为46m，塔的基部上将会有仅仅1mm厚的一条薄薄的金带，代表人类在地球上的时间。

Temple to Perspective

_ Tom Greenall+Jordan Hodgson

While writing *"Religion for Atheists"*, Alain de Botton came up with a variety of practical ways in which we may be able to change the secular modern world with insights drawn from religions.

He proposes that atheists should learn to copy the major religions and put up temples. He also suggests that atheists band together and build a network of new temples in cities across the land.

The proposal is part of a wider argument de Botton makes that atheists should stop being merely negative about religion and engage with what people actually enjoy about religion. They should then copy it – simply without the God-bit.

De Botton suggests that atheists like Richard Dawkins won't ever convince people that atheism is an attractive way of looking at life until they provide them with the sort of rituals, buildings, communities and works of art that religions have always used.

De Botton has taken the first step to making the Temple project a reality. He has come

together with Tom Greenall Architects and Jordan Hodgson to design the first of what will ideally be a chain of temples designed by different architects.

This first structure is called a Temple to Perspective and is designed to get people to recover a sense of perspective on their lives, in the midst of the multiple stimuli and clutter of the modern world.

The Temple to Perspective, which has been designed to be put up in the City of London, will be a black tower, whose structure represents the age of the earth, each centimetre equating to 1 million years. Measuring 46 metres in all, the tower will feature, at its base, a tiny band of gold a mere millimetre thick, standing for mankind's time on earth.

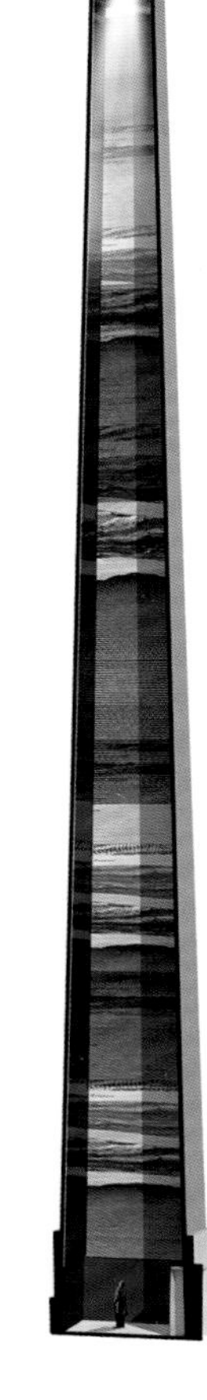

美国国家广场设计竞赛

美国国家广场是美国最受欢迎的国家公园。它为每一个美国人和国际游客提供了与美国的历史和英雄接触的机会。该公园是美国最知名的纪念碑和纪念馆的所在地，纪念碑和纪念馆用于纪念重要历史事件、民族英雄和民主理想。人们都十分喜爱这座公园。它每年接待2500万游客，举办3000场周年庆典活动。然而，建造这个占地2.8km²的公园不只是为了用于此用途，它一直没有获得足够的资源进行重建和维护，以发挥其在美国城市结构中不可替代的作用。因此，这个美国最重要的市民空间和象征空间将会被进行细致的修复和翻新，以使其永久发挥非常高的使用水平。所有用户的需求可以以一种富有吸引力的、方便的、高品质的、节能和可持续发展的方式得以满足。

美国国家广场设计竞赛使建筑师和景观设计师面临着挑战，他们要为满足广场的重建需求设计出具有创意的、可持续发展的解决方案。美国国家广场的规划着重关注三个地点：联合广场、在Sylvan剧院处的华盛顿纪念碑和宪法花园。

本次竞赛分为三个阶段，其中专业人士组成的陪审团将为美国国家广场规划中的三个需要重新设计的地点分别选择一支设计团队。这个过程包括：(1) 组合评估，为每个地点选择八名潜在的首席设计师；(2) 团队面谈，为每个地点选择五支潜在的设计团队；(3) 设计竞赛，为每个地点选择一个设计。六月初，美国国家广场信托机构已宣布选出了三支优胜团队，为美国“前院”被忽视的地点进行重新设计。

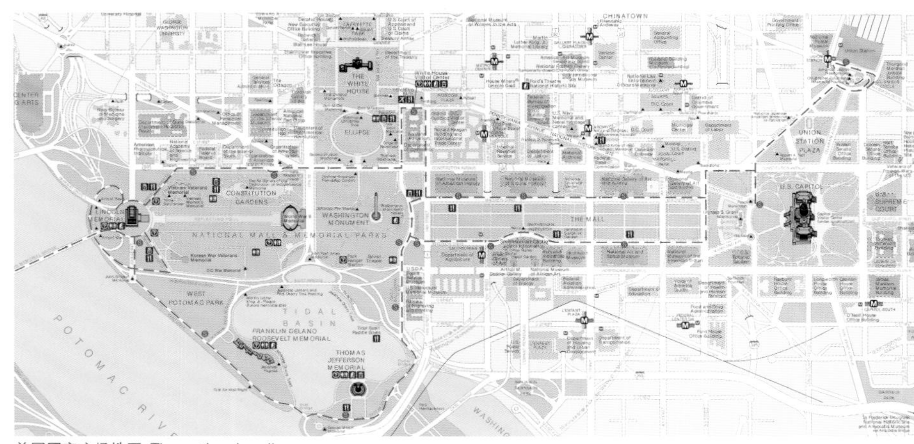

美国国家广场地图 The national mall map

总平面图 context plan

National Mall Design Competition

The National Mall is America's most popular national park. It provides every American and international visitor a chance to connect with America's history and heroes. This park is home to some of America's most recognizable monuments and memorials, all of which honor important historical events, national heroes, and democratic ideals. This park has been loved to death. It hosts 25 million annual visitors and 3,000 annual permitted events. However, this 2.8km² park was not built to withstand this level of use and has not received adequate resources to be restored and maintained to a level befitting its role as an irreplaceable piece of American fabric. Therefore, the premiere civic and symbolic space in USA, will be respectfully rehabilitated and refurbished so that very high levels of use can be perpetuated. The needs of all users can be met in an attractive, convenient, high-quality, energy-efficient and sustainable manner.

The National Mall Design Competition challenges architects and landscape architects to devise creative, sustainable solutions to the Mall's restoration needs. It focuses on three locations identified in the National Mall Plan: Union Square, Washington Monument Grounds at Sylvan Theatre, and Constitution Gardens.

This Competition is a three-stage, through which a jury of esteemed professionals will select a design team for each of three redesign sites identified by the National Mall Plan. The process includes (1) portfolio evaluations to select up to eight potential lead designers for each site, (2) team interviews to select up to five potential design teams for each site, and (3) a design competition to select a design for each site. Early in June, the Trust for the National Mall has announced the three winning teams selected to redesign the neglected sites of America's front yard.

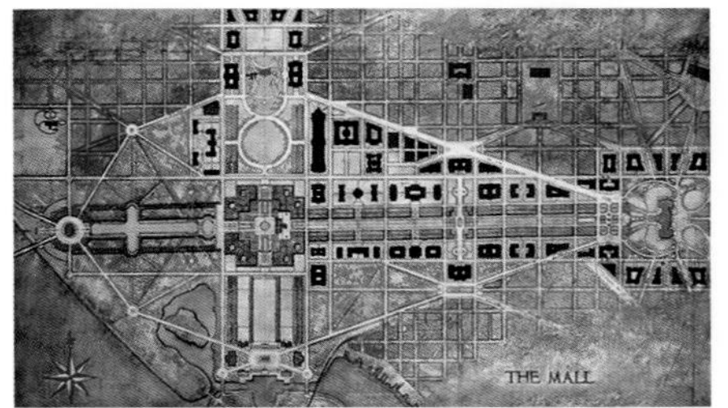

McMillan为华盛顿Source剧院所做的规划，1901年
The McMillan plan for Washington Source, 1901

照片提供：©Prince Roy
联合广场
Union Square

照片提供：©Alex Boykov
朝向西面的美国国家广场，2010年
The national mall looking to the west, 2010

照片提供：©Prince Roy
宪法花园
Constitution Gardens

Sylvan剧院处的华盛顿纪念碑
Washington Monument Grounds at Sylvan Theatre_OLIN+Weiss/Manfredi

华盛顿纪念碑位于美国国家广场中心，从数千米外即可见，是美国民族哲学层面的指南针，是永恒的精神象征，这种精神正是美国建立的基础。焕然一新的Sylvan剧院和Sylvan丛林从纪念碑这个中央地标处向外扩展，体现了莎士比亚森林给人带来的惊喜和神奇，莎士比亚森林是近一个世纪前原来剧院名称的灵感来源。

Sylvan丛林面向一系列设施，这些设施共同界定了一个新的表演景观。一片新生的枝繁叶茂的树冠和梯级草坪界定了圆形露天剧场，这里以华盛顿纪念碑的壮丽景象为背景，可进行一系列的表演和活动。上升的地貌重新定位了剧院的朝向，使其可以看到广场，隐藏了交通和旅游巴士线，并形成了连接潮汐盆地的新行人通道。最上面的观景台是种植了树木的"阳台"，它向下延伸，穿过道路，一直延伸到水边，连接了两个非凡但互不相连的景观——潮汐盆地和广场。

Sylvan馆是Sylvan丛林的扩建建筑。它能够举办即兴表演，并为到广场观光的游客提供全天开放的咖啡厅和多功能场地。该馆沿着一条向南通往独立大道的小径弯曲，在独立大道处有一个新的展览中庭，是为旅游巴士准备的到达广场和乘降点。Sylvan馆还能看到广场和各种表演的全景。

新的圆形露天剧场地形有利于各种规模的演出。咖啡厅旁细长的露天剧场可用于各种私密的和自发性的活动。

Sylvan丛林以新的表演景观复兴了纪念碑广场，明晰了白宫、华盛顿纪念碑和杰弗逊纪念堂之间的视觉联系，在广场文化景观和潮汐盆地之间形成了新的物理连接，最重要的是，为美国最显眼的中心舞台创建了一个经过改造的背景环境。

At the heart of the National Mall and visible for miles, the Washington Monument is the literal and philosophical compass for America – a timeless symbol of the spirit upon which America was founded. Extending from this central landmark, the rejuvenated Sylvan Theatre and Sylvan Grove embody the surprise and magic of the Shakespearean forest that inspired the name of the original theatre nearly a century ago.

The Sylvan Grove is oriented around a sequence of settings that together define a new performance landscape. A new wooded canopy and terraced lawn define the amphitheatre, where a wide range of performances and events are seen against the stunning backdrop of the Washington Monument. This ascending landform re-orients the theatre to views of the mall, conceals traffic and lines of tour buses, and creates new pedestrian connections to the Tidal Basin. The uppermost viewing terrace, a tree lined "balcony", continues across the roadways to descend to the water's edge, linking the Tidal Basin and the Mall, two extraordinary but disconnected landscapes.

The Sylvan Pavilion is an extension of the Sylvan Grove. It hosts impromptu performances, and offers an all-weather café and multi-use destination for visitors to the Mall. The pavilion bends along a pathway south to Independence Avenue, where a new exhibition atrium frames an arrival plaza and drop-off for tour buses. It also offers panoramic views of the Mall and performances.

The new amphitheatre topography supports performances of all scales. The slender amphitheatre adjacent to the café supports intimate and spontaneous events.

The Sylvan Grove reinvigorates the Monument Grounds with new landscapes of performance, clarifies the visual connections between the White House, the Washington Monument, and the Jefferson Memorial, provides new physical connections between the cultural landscape of the Mall and the Tidal Basin, and most importantly, creates a transformed setting for America's most visible ceneral stage.

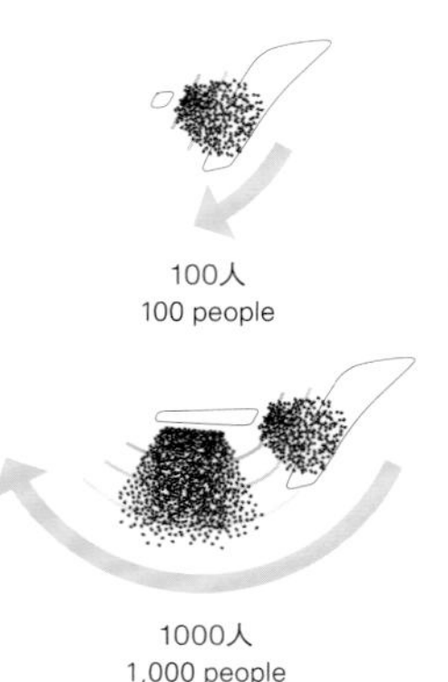

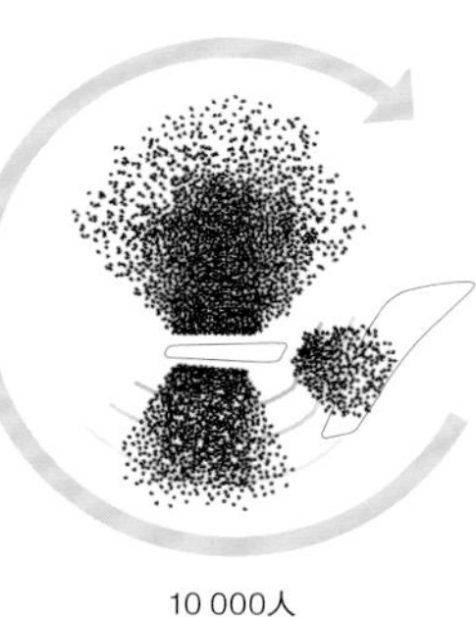

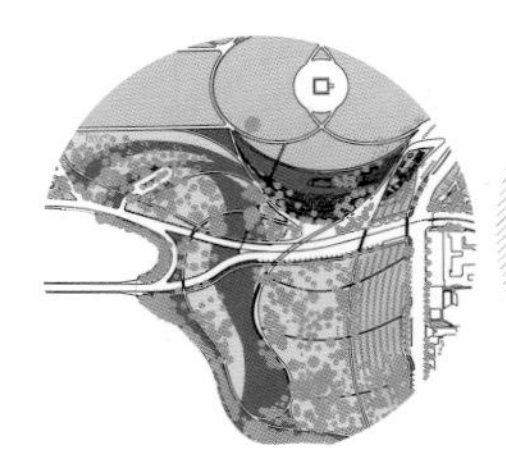

508,550 SF
PROPOSED
LAWN

1,322,795 SF
existing lawn

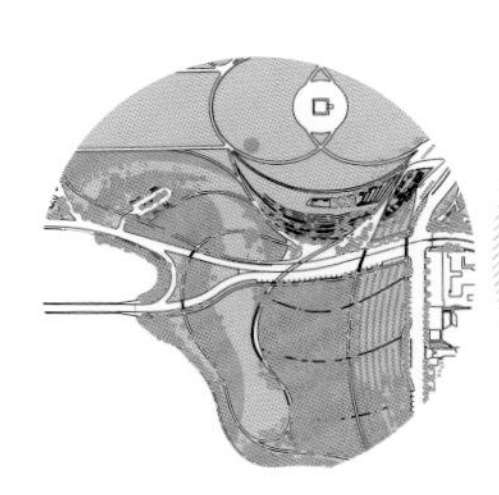

1,322,795 SF
existing lawn

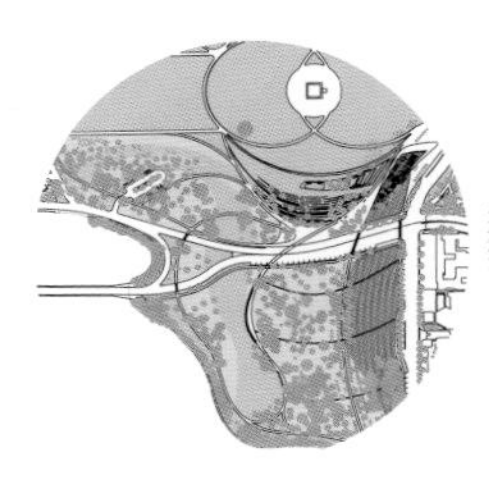

1,322,795 SF
existing lawn

联合广场

Union Square_Gustafson Guthrie Nichol+Davis Brody Bond

联合广场是纪念美国某些最重要的共享记忆的设施。它是美国人民在民族危机和庆祝时刻的聚集地，是一个表现美国人民团结的标志性公共空间。

今天，联合广场是一个充满水和草坪的广阔的、未修饰的平面区域。现在有了用其他不太正式的空间、特点、纹理来覆盖和丰富这个广场的机会，特点和纹理是根据此处的潜在自然地貌而设计的，每天穿过城市结构的运动模式以及全年来这里的人们的多种需要与愿望也使广场的设计更加丰富。

联合广场设计规模巨大，给游客留下了深刻印象，设计还为一系列的体验和声音提供了舒适场所。大型集会的主要焦点将是一个由倒影池和格兰特纪念碑定义的大型中央空间。这个水池的水只有浅浅的一层，下面是一层铺砌表面，可以排光所有的水，形成一系列不同大小的集会空间。铺砌道路沿对角线方向穿过水池表面，可以满足穿越这个大水池的行人的需要，但也允许人们在水面上玩耍，使这个正式空间充满活力和人性化。

景观形成了一系列个性鲜明的室外空间，高度的微妙变化形成了梯级和低矮的挡土墙，这些挡土墙可以为人们提供座位，还能让人们欣赏到这些室外空间。

广场利用种植原有成熟的树木对内部空间进行了进一步定义。建筑师选用了特定的材料以突出组成空间的不同特征，同时创造持久的室外空间，这些室外空间可以在该项目的生命周期内得到高效维护。

联合广场由独特的单个元素组成，具有统一的特点，它的建成来源于丰富的空间纹理排列和人类经验，然而却获得了象征性作用在美国所必需的独有的特点和力量。

Union Square is the setting for some of America's most powerful shared memories. It is the ground on which Americans gather at moments of national crisis and celebration and an iconic public space that expresses their unity.
Today, Union square is a broad and unmodulated plane of water and lawn. There is the opportunity now to overlay and enrich this plaza with other less formal spaces, features and textures that respond to the underlying natural landform of the site, the day-to-day patterns of movement through the city fabric and the diverse needs and desires of the people who populate this place throughout the year.
This design for Union Square engages visitors with an impressive monumental scale, but it also provides comfortable places

公共广场+水池 public plaza + pool

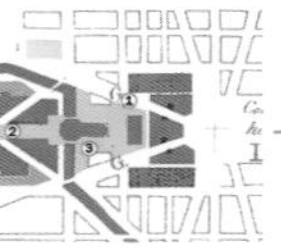

公共广场+水池
public plaza + pool
L'Enfant plan_1791

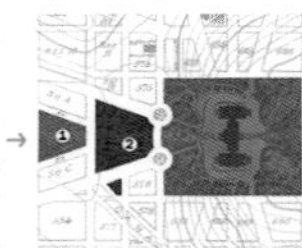

公共花园+植物园
public + botanical garden
report of the commisioners, distinct columbia_1891 from Andrew Jackson Downing design_1851~1852

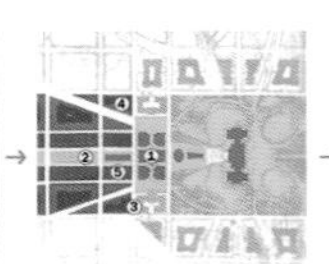

公共广场
public square
The McMillan plan_1901

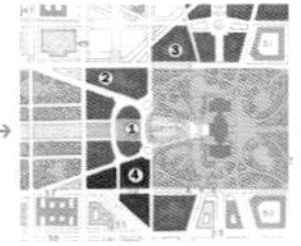

框架界定的轴线+纪念碑
framed axix + monument
McMillan plan adaptation by Frederick Law Olmsted, Jr_1932~1936

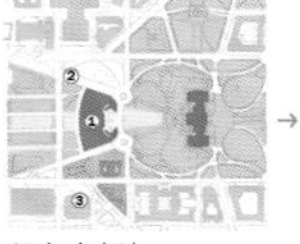

纪念碑水池
monument pool
existing condition, from 1976 development plan by Skidmore Owens and Merrill with Dan Kiley

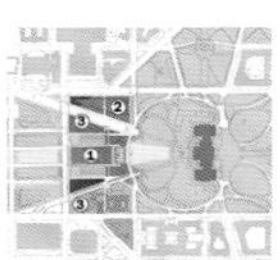

当前设计：统一的场地
current proposal: unified ground
proposed condition

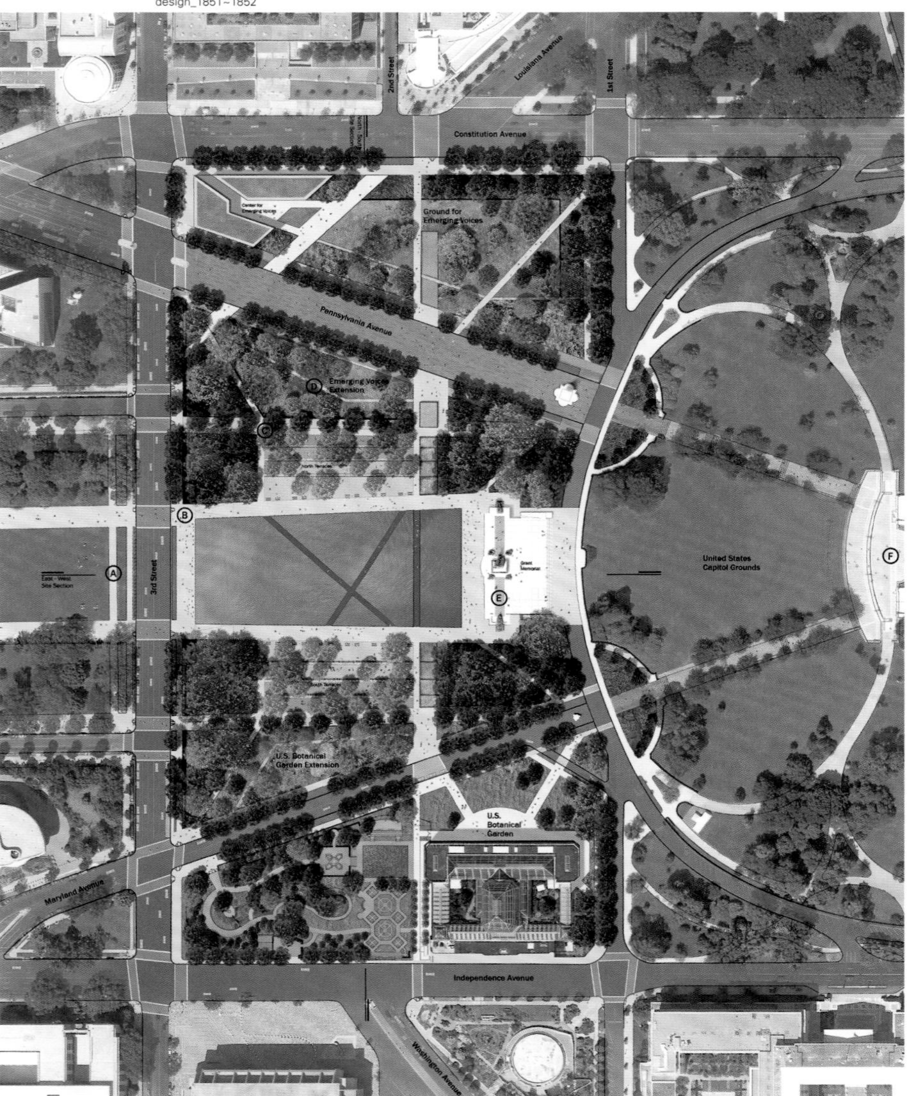

for a range of experiences and voices. The major focus for large assemblies will be a grand central space defined by the reflecting pool and Grant Memorial. Consisting of a shallow sheet of water over a paved surface, this pool can be drained to create a range of differently sized assembly spaces. Paved paths running diagonally at the surface of this pool serve the need of pedestrians to traverse this large water feature, but they also allow people to inhabit the water's surface, enlivening and humanizing this formal space.
The landscape forms a series of outdoor rooms of distinct character; subtle changes in grade are used to form terraces and low retaining wells that provide seating surfaces and views into these rooms. Planting further defines the spaces within, utilizing the existing mature trees on the site. Materials have been selected to reinforce the different identities of the constituent spaces, as well as to create durable outdoor spaces that can be efficiently maintained through the life cycle of the project.
A unified composition, composed of distinct and individual pieces, Union Square is built from a richly textured array of spaces and human experiences, yet achieves the singular identity and power necessary for its symbolic role in America.

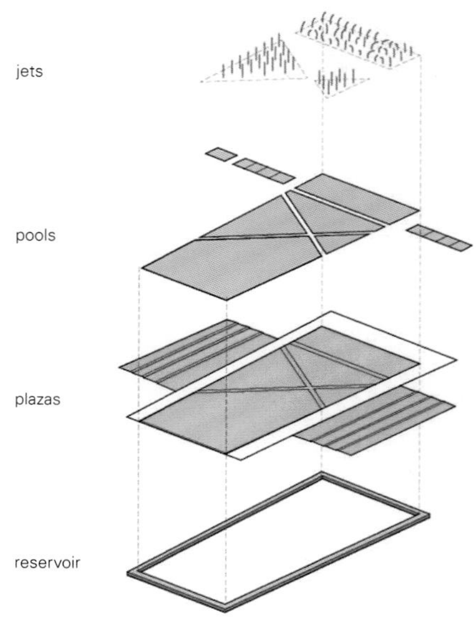

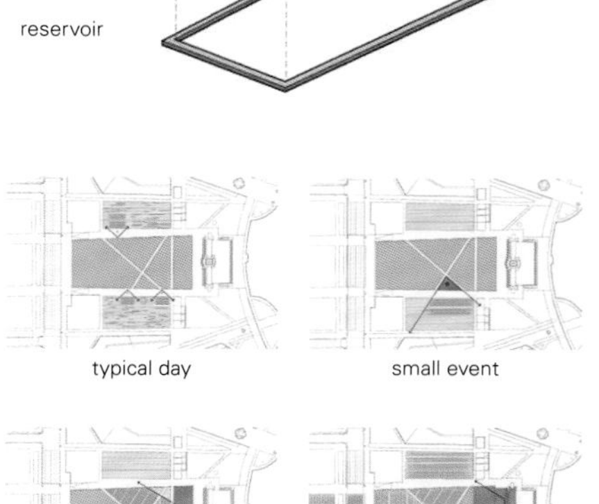
typical day
small event

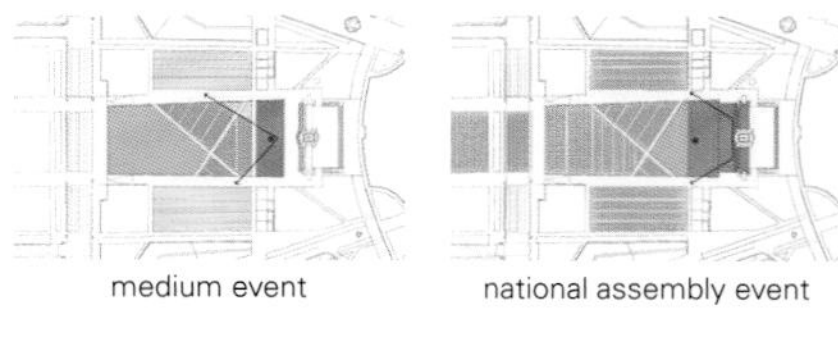
medium event
national assembly event

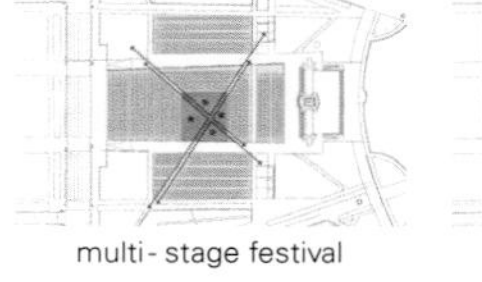
multi- stage festival

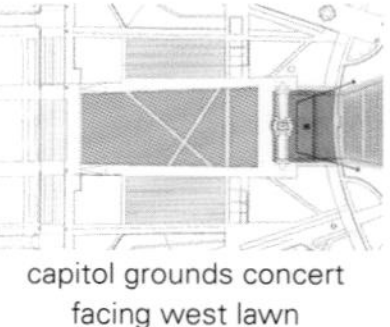
capitol grounds concert facing west lawn

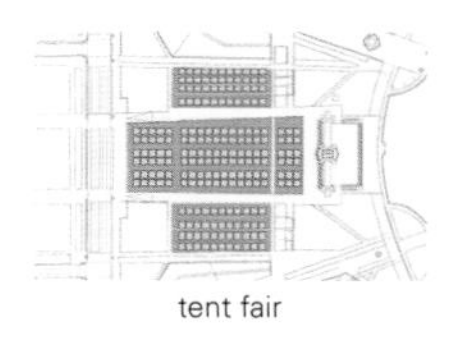
special stage performance
tent fair

表现美国民族特点
to express national identity

identities

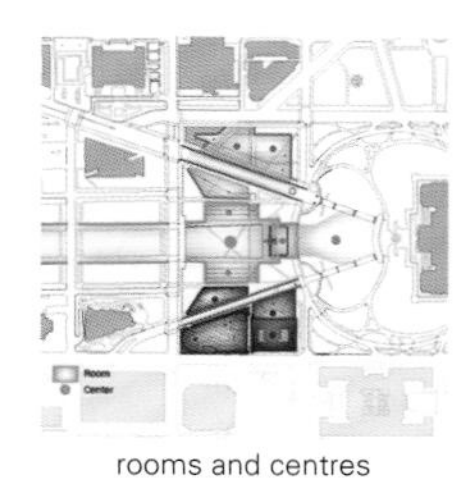
rooms and centres

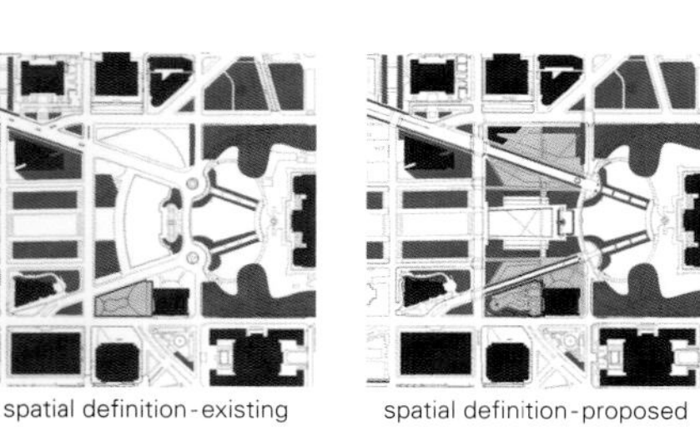
spatial definition - existing
spatial definition - proposed

遇到差异
encounter difference

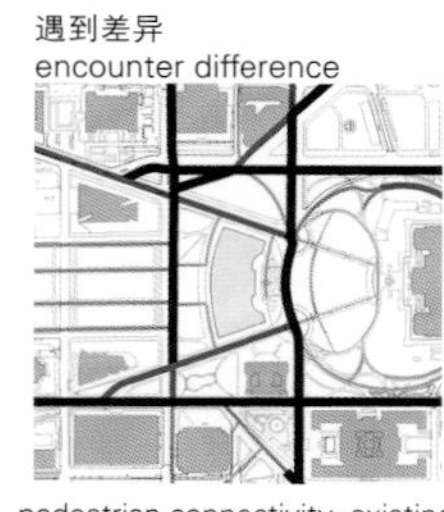
pedestrian connectivity - existing

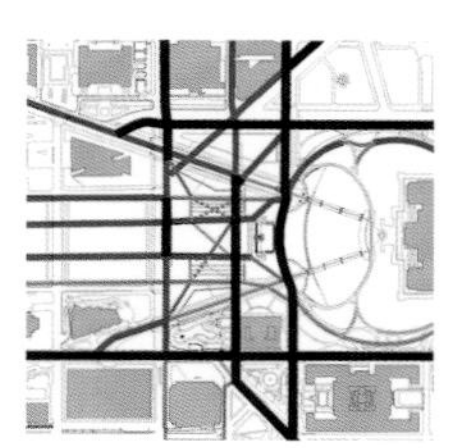
pedestrian connectivity - proposed

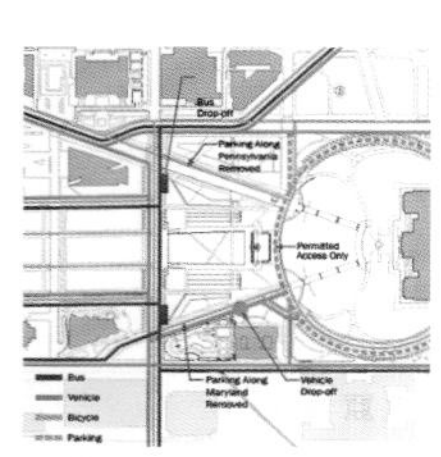
vehicular access and routes

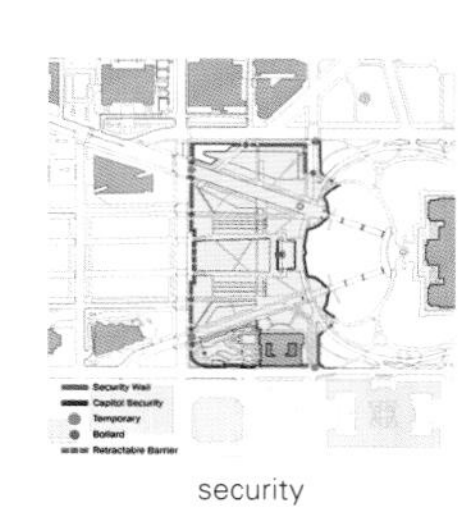
security

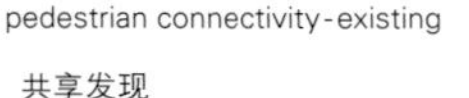
共享发现
share discoveries

paths and materials

tree canopy

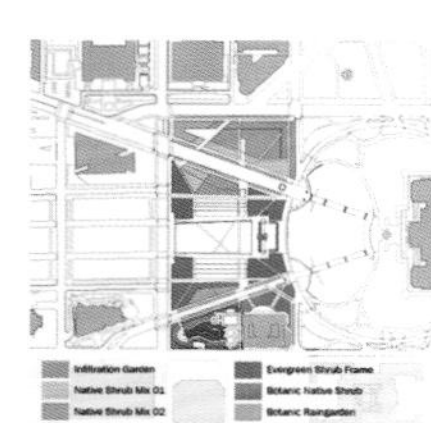
planting identities

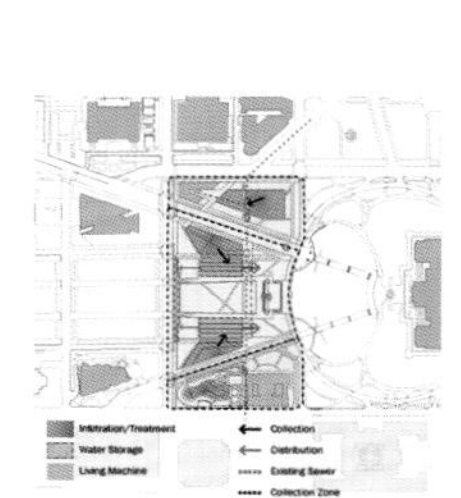
site hydrology

宪法花园

Constitution Gardens_Rogers Marvel Architects+Peter Walker and Partners

宪法花园的设计尊重原有的规划，旨在提高其非凡的特性，提高其形式美，明确其目的，刺激人们所有的感官。该设计为对基地进行生态和社会再生建立了一个动态框架，尽管年久失修，但这里仍然是一个重要的值得保存的现代景观。

宪法花园将作为一个美观的生态园进行重建。通过对规模、形态和景观各组成部分功能的精心改进，建筑师将创造一个美好的、具有生态生产力的集成环境。建筑师将把湖改造成水基础设施的一个关键部分，水基础设施能减少雨水的破坏性影响，同时创造可再利用的水资源。宪法花园仍将是一个理想化的自然景观，但它会打破生态与美观的对立状态，形成一个二者兼顾的景观。

湖东端的新展馆将同时作为宪法花园的门槛、锚固点、活动之间的连接。展馆从规模到体量和材料，都融入到了景观之中，毫不含糊地吸引着人们并保护着景观。新的草地圆形露天剧场可容纳数千观众，也是日常野餐和休闲的好去处。

宪法花园将保留其作为乡村设施的初衷，但无论在什么季节，不管是白天还是夜晚，都能提供新的用途。

The design for Constitution Gardens respects its original plan and seeks to heighten its unusual nature, enhance the beauty of its forms, clarify its purpose, envelop all the senses. This proposal establishes a dynamic framework for ecologically and socially regenerating a site that, though in disrepair, is an important modern landscape worth preserving.
Constitution Gardens will be rebuilt as an aesthetic ecology. By carefully refining the scale, shape, and function of each component of the landscape, the architects will create a beautiful integrated environment which is ecologically productive. The architects will transform the lake into a critical piece of water infrastructure that reduces the damaging impacts of stormwater while creating a source of water for reuse. This gardens will still be an idealized version of nature, but it will break out of the false dichotomy of ecological versus aesthetic to create a landscape that is both.
A new pavilion at the east end of the lake will operate simultaneously as a threshold, an anchor point, and a nexus of activity for Constitution Gardens. From scale to massing and materials, the pavilion is designed to integrate into the landscape, to entice and protect without obscuring. A new grass amphitheatre can accommodate an audience up to several thousand and also make a great spot for everyday picnicking and relaxation.
Constitution Gardens will retain its original purpose as a pastoral setting, but offer new uses and events, in all seasons, day and night.

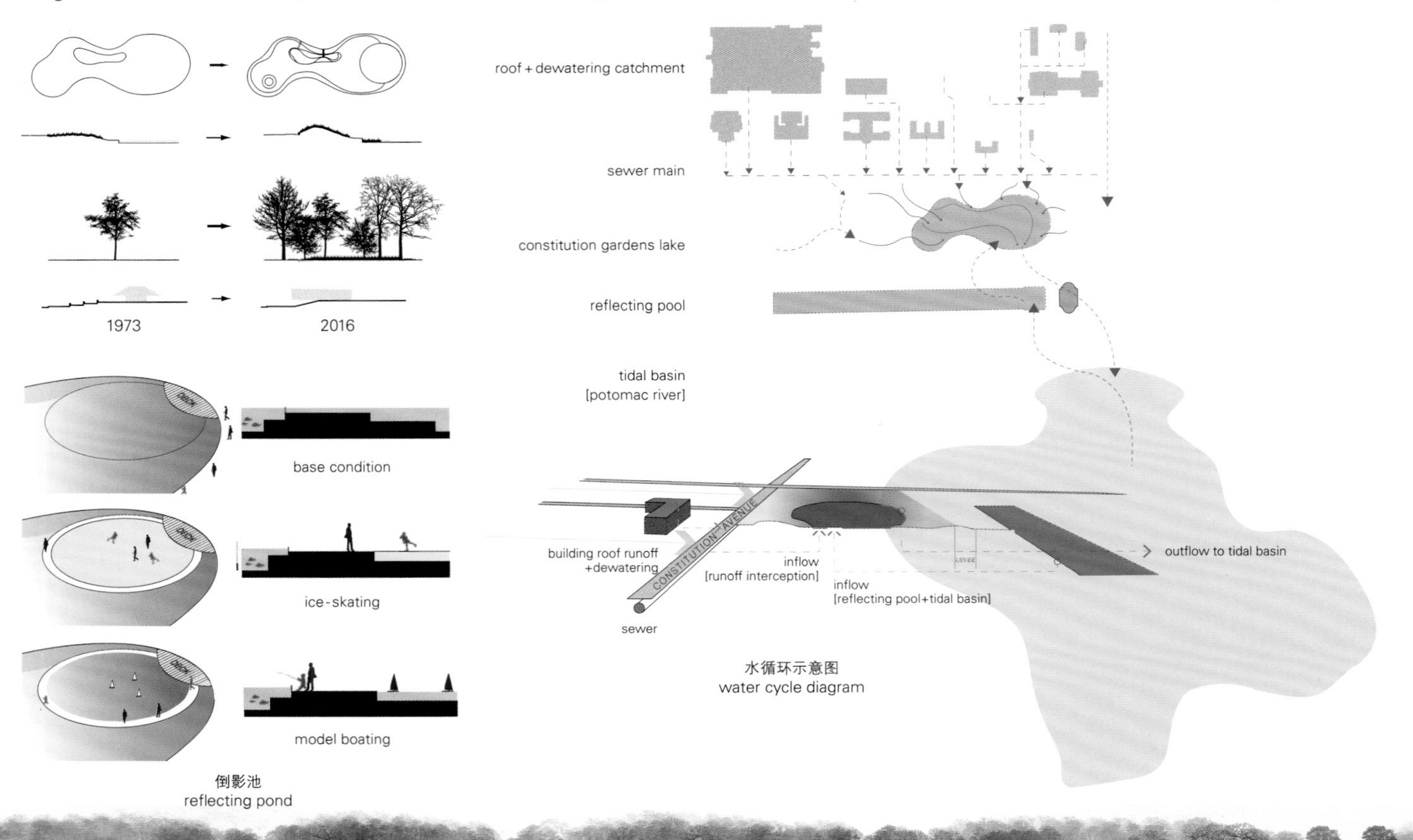

倒影池
reflecting pond

水循环示意图
water cycle diagram

The landform will be subtly made more elegant with higher elevations toward Constitution Avenue and Seventeenth Street providing separation and locations for viewing the park.

地形 topography

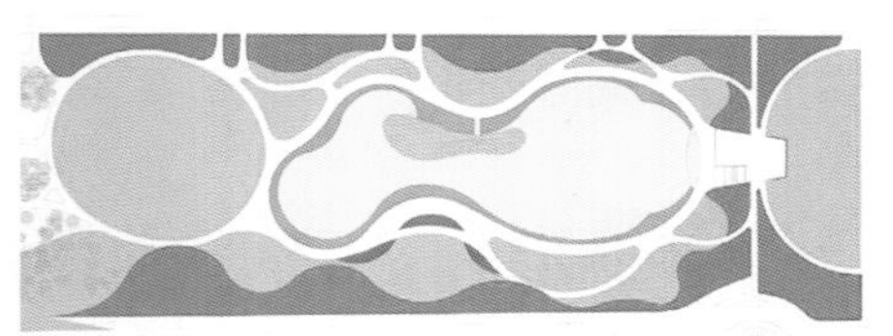

■ woodland ■ lawn + meadow ■ lowland garden
■ upland garden ■ aquatic shelf

下层植被 understorey planting

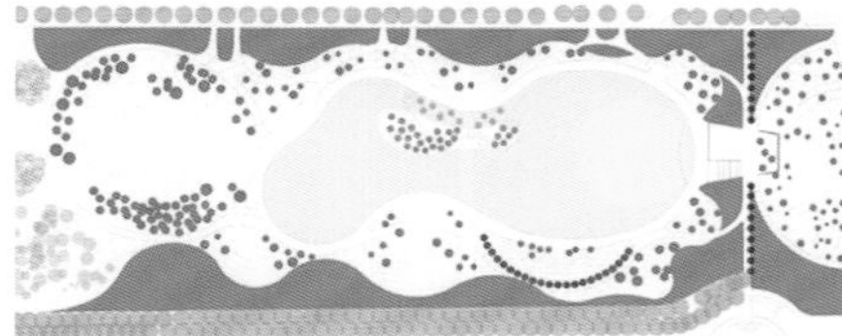

■ woods ● specimen tree • magnolia • willow
● existing elm • formal row

树木 tree planting

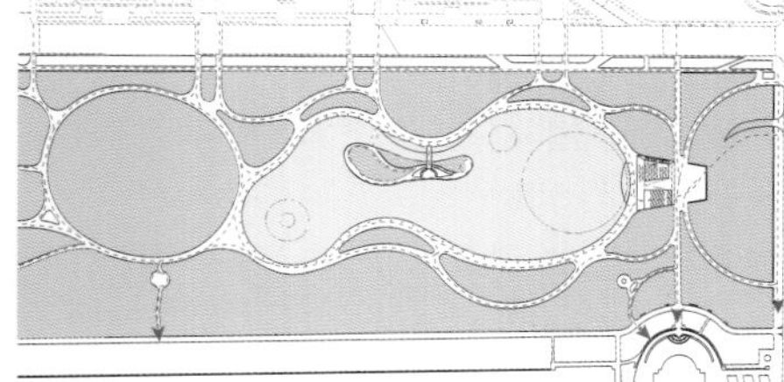

wider paths surrounding the lake as well as smoother pedestrian connections will encourage use of the garden and provide better access to other memorials and monuments on the mall

交通流线 circulation

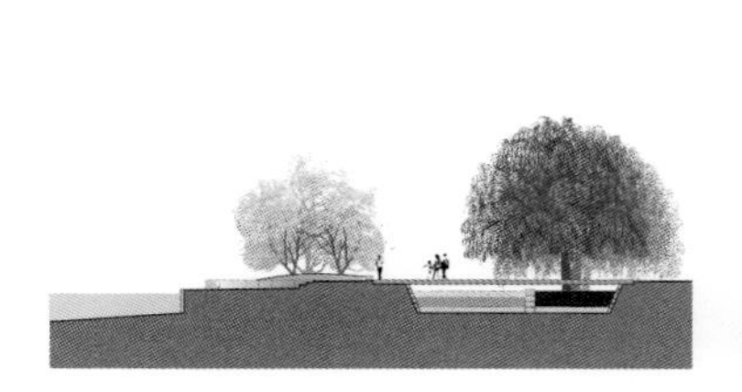

阿尔梅勒奥斯特沃尔德的城市规划

阿尔梅勒东侧的开发区域是该市"大飞跃"的一部分，是城市和开拓地农业景观之间的过渡地带。它为人们的生活、工作和娱乐提供了具有绿色和乡村特色的空间。该开发区域将具有非常低的建筑密度，从而反衬出城市西侧的高度城市化发展。

阿尔梅勒奥斯特沃尔德为单独和总体设计提供了最大的自由度，将实现较长时间的有机增长。为了促进阿尔梅勒市可持续发展目标的实现，50%的面积将被用于发展城市农业，为城市生产与当地生产规模相同的产品，并保持当前的农业特色。该开发战略是以单独设计为基础的，将会整合绿地和城市规划，将大规模开拓地转变为更加具有差异性的景观。

该开发项目的每个设计者都将遵循一套原则，但以一个主要的原则为基础：你可以做你想要做的一切，但你必须自己规划好一切。设计者将根据预先确定的分支项目开发他们的试点项目的所有元素，分支项目包括一段公共道路、绿地、水缓冲区域和城市农业空间等。每个试点项目周围都设有连续"路径"和为所有建成项目所做的从道路处回退的设计，这将保证形成可渗透的、连续的绿色景观。

阿尔梅勒奥斯特沃尔德的开发战略将引进一种全新的城市模型，这种模型完全面向用户，更加灵活，从而可将这个区域改造成一个具有多样性、实验性和充满惊喜的区域。它将阿尔梅勒的私人委托开发策略推向了集体城市规划的更高水平。

Urban Plan Almere Oosterwold

_ MVRDV

The development of the eastside of Almere, part of the "quantum leap" of the city, is the transition zone between the city and the agricultural landscape of the polders. It offers space for living, working and recreation with a green and rural character. The area will be developed with a very low building density, and will thus counter the highly urban developments on the west side of the city.

Almere Oosterwold offers maximum freedom for individual and collective initiatives and will grow organically over a longer period of time. To contribute to the city's sustainability goals, 50% of the area will be used as urban agriculture, producing products for the city on a local scale and maintaining the current agricultural character. This development strategy based on individual initiatives will transform the large-scale polders into a more differentiated landscape, integrating green and urban programs.

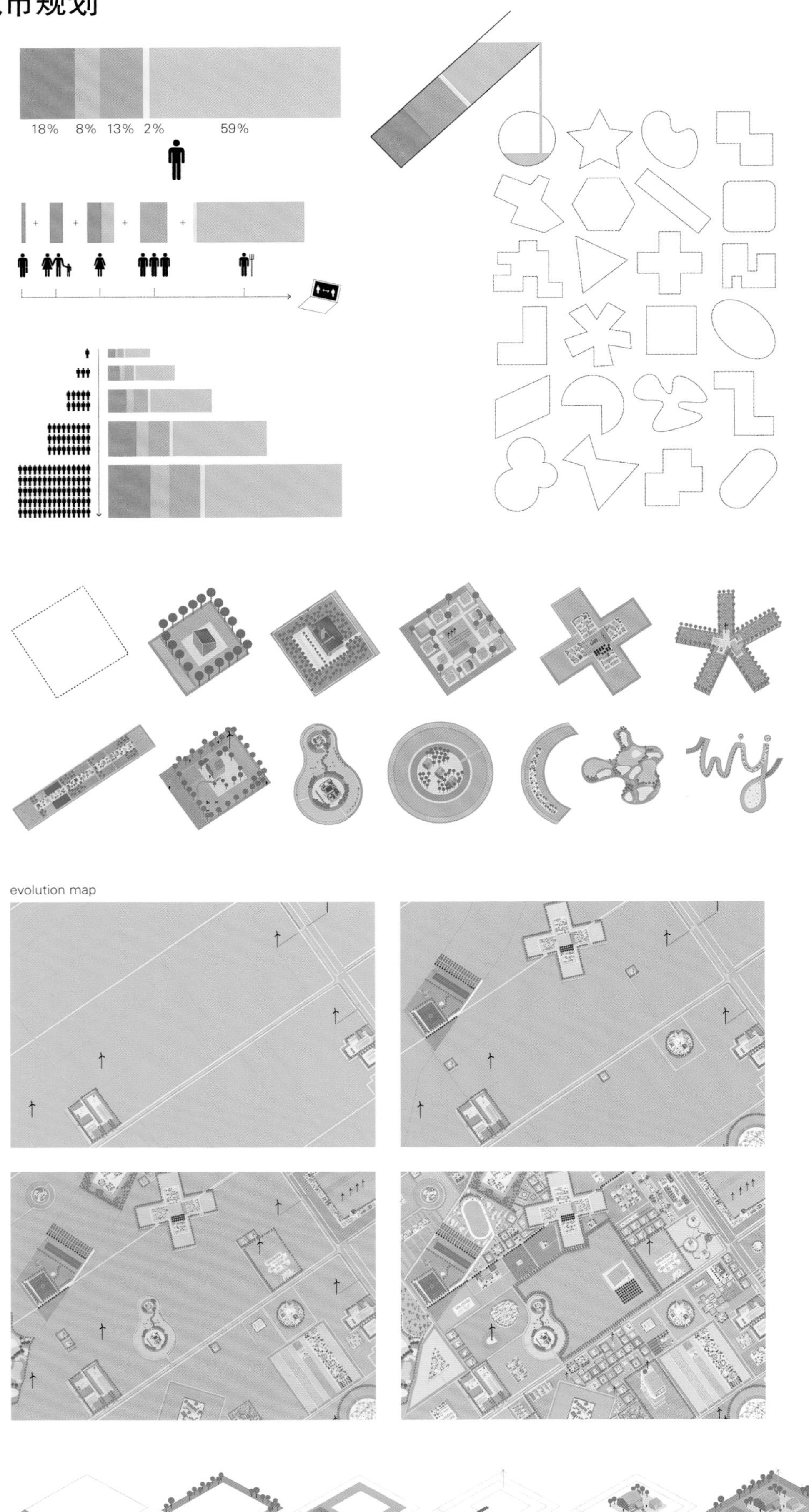

The development will be guided by a set of principles for each initiator, based on one main principle: you are able to do everything you want, but you have to arrange everything yourselves. The initiators will develop all elements of their plot according to a predefined programmatic division, which includes a piece of public road, green space, water buffer and space for urban agriculture. A continuous accessible "path" around each plot and a setback from the roads for all built program will guarantee a permeable and continuous green landscape.

This development strategy for Almere Oosterwold will introduce a radical new urban model which is completely user-oriented, more flexible, and will thus transform the area into a district that breathes diversity, experiment and surprise. It takes Almere's well developed strategy of private commissioning to the next level of collective urban planning.

项目名称：Almere Oosterwold, NL
地点：Almere Oosterwold, the Netherlands
建筑师：MVRDV
设计团队：Winy Maas, Jacob van Rijs, Nathalie de Vries, Jeroen Zuidgeest, Klaas Hofman, Chiara Quinzii, Mick van Gemert, Sara Bjelke, Jonathan Telkamp, Maarten Haspels, Wing Yun
景观设计：DLG; Niels Hofstra
基础设施、能源、卫生设备：Grontmij; Alex Hekman, Martin de Jonge, Jasper Groebe
城市农业：Wageningen University, Jan Eelco Jansma
艺术效果图：MVRDV
甲方：Werkmaatschappij Almere Oosterwold / Municipality of Almere
用地面积：4.300ha
规划：15,000 dwellings, 26,000 jobs, 135ha business, 20ha bvo offices, facilities, 400ha new landscape
设计时间：2011.4 — on going 竣工时间：2013

埃塞俄比亚国家体育场与体育村

为热爱足球和田径运动的埃塞俄比亚人建造的埃塞俄比亚新国家体育场将有一个新的满足国际足联和奥运标准的60 000座体育场，这个体育场位于亚的斯亚贝巴，是结合了当地特色与新技术的设计竞赛获胜方案。

LAVA（视觉建筑实验室）和Designsport一同与埃塞俄比亚当地公司JDAW合作，赢得了埃塞俄比亚联邦体育委员会举行的国家体育场和体育村的国际建筑设计竞赛。

LAVA回归了体育场设计的原点，设计了下沉的运动场，周围环绕着用挖掘的土方建成的大看台。这种人造大坑是对现有地形进行巧妙重塑的结果，这样的设计形成了有效的空间，优化了环境性能，最大限度地减少了建设成本，并将设施融入了现有的景观中。

从上面看，体育场的结构形式很像咖啡豆，咖啡豆是埃塞俄比亚的主要收入来源，也像是"母亲的子宫"，是约320万年前第一批人类中的一员——露西的骨架。

该体育场的屋顶是一层智能薄膜，是漂浮在地面景观之上的轻质张拉结构，很像广阔的埃塞俄比亚天空中的一片云。

总体规划包括建造用于举办国际足联赛事、体育赛事、音乐会、宗教活动和民族节日庆典的国际奥委会标准体育场，还要建造一个体育村，包括室内和室外的水上娱乐中心、室外球场、运动大厅和场馆、为联邦体育委员会设置的宿舍和总部。酒店、零售和商业区域将确保该地区一年四季充满活力。

结构构造和运动流线是总体规划的基本概念。周围Entoto山的壮丽美景是设计的背景，设计与该地区的火山地质相呼应。和缓起伏的城市公园沿着大坑的线条变化，给人一种连续的空间体验，能策略性地平衡运动流线、气候、体验与效率。中央广场构成该项目的核心，是连接所有区域的山脊。

太阳能供电的巨型雨伞提供了遮阳之处，而由行人激活的灯光与水景以地面裂缝的形式出现，提供给人们寻找和创造动画艺术作品的方式。

该体育场预计于2014年开始建造。

National Stadium and Sports Village_LAVA+Designsport+JDAW

New national stadium for football-loving Ethiopians and athletics-loving Ethiopians will have a new FIFA and Olympic-standard 60,000 seats stadium in Addis Ababa thanks to a competition winning

design combining local identity with new technology.
LAVA, the Laboratory for Visionary Architecture, and Designsport collaborated with local Ethiopian firm JDAW to win the international architecture competition for a national stadium and sports village, held by the Federal Sport Commission, Ethiopia.
LAVA went back to the very origin of stadium design with a sunken arena surrounded by grandstands formed from excavated material. This man made crater is a clever remodelling of the existing terrain and generates efficient spaces, optimises environmental performance, minimises construction costs and integrates facilities

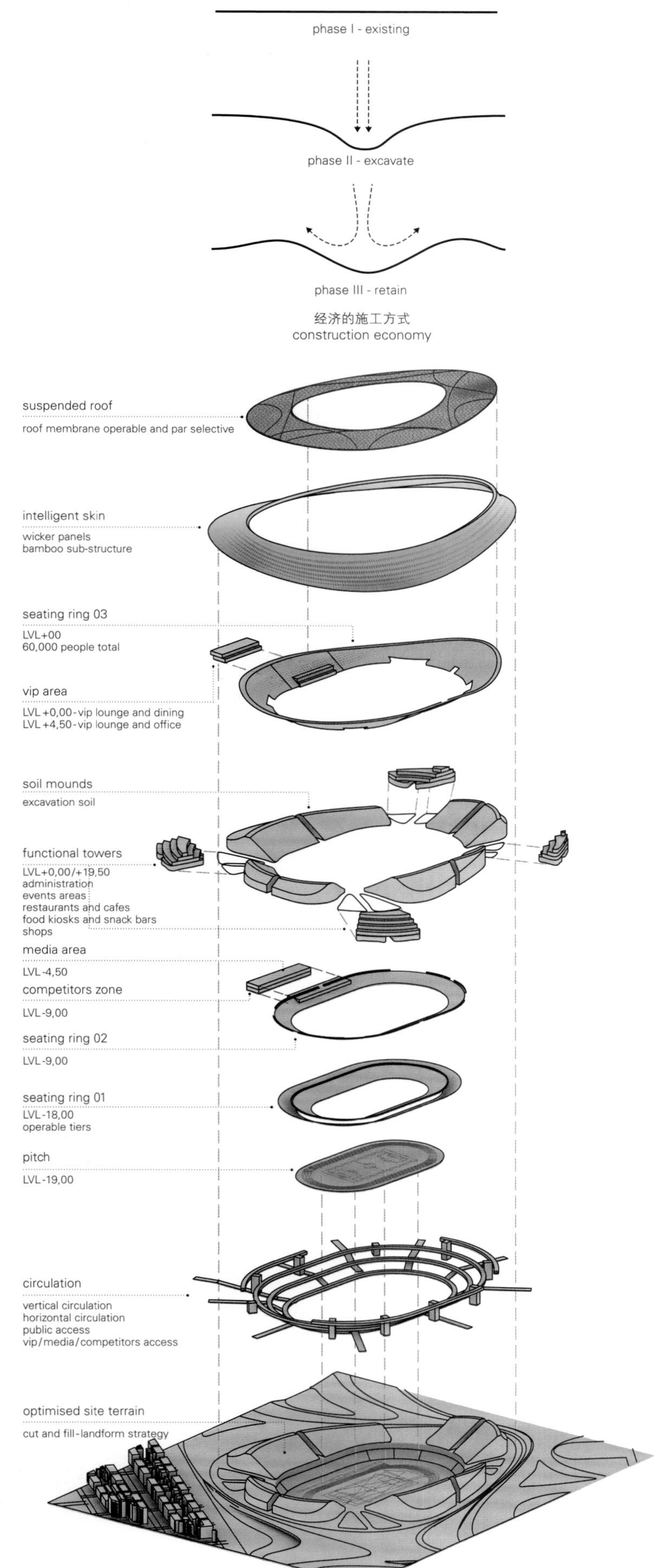

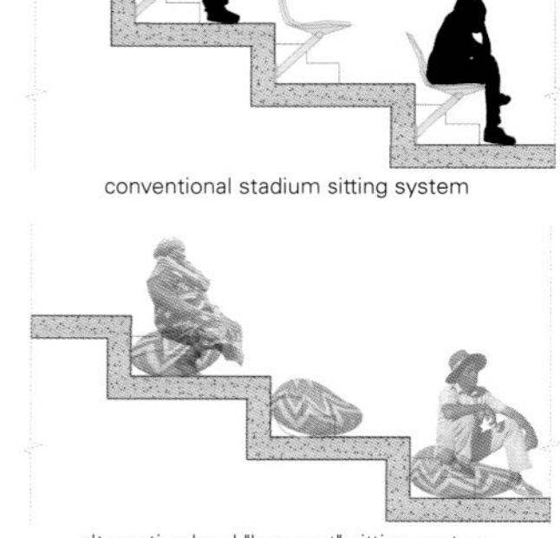

灵活的座位系统
flexible seating system

within the existing landscape.
The form of the stadium structure seen from the top view also recalls coffee beans, the main source of income in Ethiopia and the "Mother womb", the skeleton of one of the first humans, Lucy, which is about 3.2 million years old.
The roof of the stadium, an intelligent membrane, appears like a cloud on the horizon of the vast Ethiopian sky, a lightweight tensile structure floating over the formed-earth landscape.
The masterplan includes the IOC-standard stadium for FIFA matches, athletics events, concerts, religious and national festivals; and a sports village comprising indoor and outdoor aquatic centres, outdoor pitches, sports halls and arenas, dormitories and the headquarters for the Federal Sport Commission. Hospitality, retail and commercial zones will ensure that the precinct is vibrant throughout the year.
Tectonic structures and movement are the underlying concept for the masterplan. The breathtaking beauty of the surrounding Entoto Hills is the backdrop to a design that responds to the volcanic geology of the region. Gently undulating urban parkland follows the lines of the crater and is conceived as a continuous spatial experience strategically activated to balance movement, climate, experience and efficiency. A central plaza forms the heart of the project and a ridge connects all zones.
Giant solar powered umbrellas provide shade and shelter whilst pedestrian activated light and water features appear as fissures in the ground surface, providing way finding and creating animated art works.
The construction of the stadium is expected to commence in 2014.

项目名称：National Stadium and Sports Village
地点：Addis Ababa, Ethiopia
建筑师：LAVA Laboratory for Visionary Architecture - Chris Bosse, Tobias Wallisser and Alexander Rieck with Jarrod Lamshed, Angelo Ungarelli, Vivienne Ni, Paul Bart, Giulia Conti, Alessandra Moschella, Teresa Goyarrola, Manuel Caicoya, Guido Rivai
Designsport – Samantha Cotterell, Basil Kalaitzis, Irene Roccia
当地建筑师与工程师：JDAW Architects-Daniel Assefa; Mesfin Bekele, Salsawi Seyoum, Martha Hadish, Azeb Eshetu, Fikreselassie Sifir
技术与文化协调：John Shenton
甲方：The Federal Sports Commission of Ethiopia
规划：60,000 seat stadium, athletics track, aquatic centre, residential village, headquarters, Federal Sport Commission; and sports, halls and arenas
用地面积：600,000m²
施工时间：2014 — (expected)

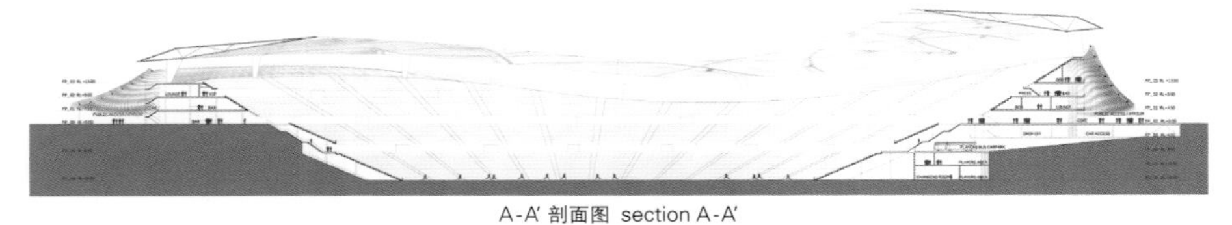

A-A' 剖面图 section A-A'

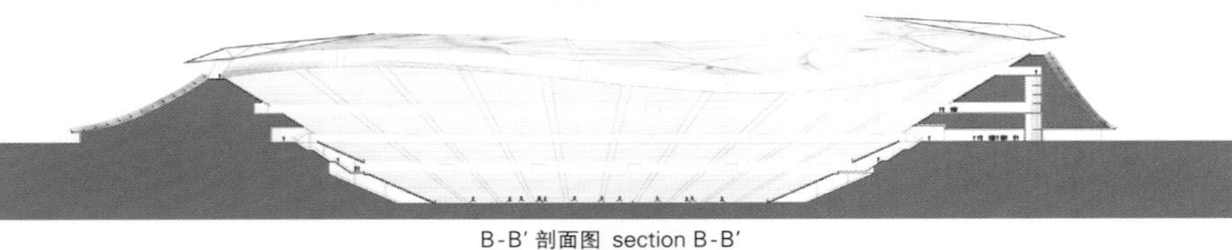

B-B' 剖面图 section B-B'

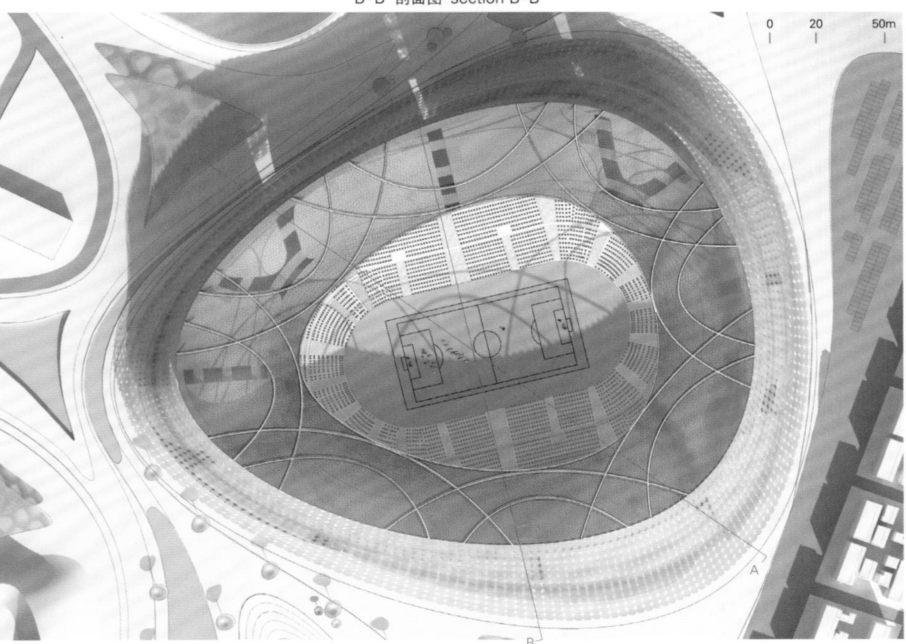

Add in the Scape

建筑入景

20世纪颠覆了现实与自然之间的普遍认同理念，物理学、哲学和艺术在理解现实和阐释现实方面都引入了全新的层次结构。空间和时间的传统欧几里德概念被彻底颠覆，随之形成的是已确立的内外之间、观察者与观察对象之间的层次结构，从而呈现出人类环境的结构。人们不再认为"内""外"的概念是固定的、无可争议的。

人与自然以及景观结构之间的关系，已经从彼此限制转变成彼此融合。人们将景观改造为故事的发生地，在这里演员和观众的角色界限变得模糊。此外，关于人类存在的可持续性的新意识使新材料的实用性大大增强，因此也形成了材料形状和特征的全新语言。

这里所讨论的项目是运用不同方法使建筑融入景观的实例。使建筑模仿自然形状从而融入景观，或者借助创造戏剧场景的方式，抑或是用类比的方法形成一种原始的内在形态，它们在自身和周围的文脉中创建了一种全新且微妙的辩证法。

Overcoming the popular identification between reality and nature, in the Twentieth Century, physics, philosophy and the arts have introduced new hierarchies in the aspect of seeing and interpreting reality. The traditional, Euclidean concepts of space and time were completely revolutionized, with them the well-established hierarchies between interior and exterior, between observer and observed, that characterized the construction of the environment of mankind. "In" and "out" ceased to be considered solid and undisputed concepts.

The relationship between man and nature, and landscape construction itself, has gone from a process of imposing to a process of incorporation. The landscape is transformed into a scene in which the roles of actor and spectator become ambiguous. Furthermore, a new awareness about sustainability of human presence has produced a great availability of new materials and, consequently, a new language of shapes and signs.

The projects discussed here are examples of different approaches to the issue of incorporation into the landscape. Incorporating the building in the landscape as mimesis of natural shapes, or by means of creation of theatrical scene, or evocating analogically ancestral inner states, they establish a new, subtle dialectics between themselves and the surrounding context.

融入景观之中
Incorporated in the Scape

融入自然，融入风景

没有融合就没有景观的存在。自然和景观呈现出一种几乎巧合的寓意，这种天真的说法算得上是老生常谈，但情况却并非如此。通过对周围环境进行改造，只有将我们自己的人造建筑融入这种环境之中，人类才会将自然从偶然变为最终的景观。

"融入"作为一个过程，并不总是与"模仿"相似。一座建筑要想融入自然，最主要的方面是建立一个全新的景观。任何融入已经被人类塑造好的景观的行为都会使该景观不断发展，从而包含一种全新的品质，这种品质既可以是正面的，也可以是负面的。

在极少数的情况下，普通人仍然能够接近自然，任何与自然的直接联系都会通过一种庄严的方式形成。这是一种纯粹的情感关系，缺少语义上的连接。此时此地，没有发展带来的影响，只有一种纯粹的审美愉悦，一种只为哲学家和喜欢沉思的旅行者保留的特权，然而现在很明显地扩展到喜欢冒险旅游的游客当中。在对自然的直接感悟中，我们作为完全脱离于自然的旁观者，为那些无法进入景观中的人们亲身感受自然。

我们与景观的关系，要比我们与自然的关系复杂得多，它涉及到一种概念关系，我们可以而且应该进入到这个场景中，而且在该场景中是作为一名导演而非演员。

城市景观或乡村景观都是人类的舞台，无论我们是在这片景观中劳作还是休闲，利用资源还是节约资源。我们可以做出自己的行动，形成自己的礼制，满足自己无论是物质还是非物质方面的需求，也可以表达自己的世界观。

观察者和观察对象

他突然大声说："这就是薛定谔想要表达的！"他非常兴奋，"Julie，还记得薛定谔客观化文章的最后一部分吗？'主体和客体是一体的。它们之间的障碍不能说是被最近的物理科学所打破，因为这种障碍根本就不存在。'"从我们讨论那篇文章开始我就一直对这种说法迷惑不解。现在考虑到你所说的，我才明白了其中的含义。在沉思中，人们可以触及真相。这就是你所谈到的这种融合，也可以说是万物合一。没有主体也没有客体存在。只有当一个人的思想回归到具体化的话语模式时，主体和客体的二元性才会存在一个分歧点。薛定谔一定通过沉思感受到了他所写出来的事物。一切事物开始结合在一起。[1]

20世纪，量子物理学提出一种过时的理念——不管一个人多么孤独，他都可以处于观察系统之外。观察者是系统的一部分，观察的行为会使系统发生实际变化。这一结论非常适用于人及其所处的场景之间的

Incorporation into Nature, Incorporating with the Landscape

No landscape exists in the absence of incorporation. A naïve commonplace holds that nature and the landscape assume an almost coincidental meaning, but such is not the case. It is only by incorporating our own artifacts into an environment, by manipulating surroundings, that humankind transforms nature from happenstance into the causality of landscape.

"Incorporation" as a process is not always similar to "mimesis". The primary aspect of the incorporation of an architectural object into nature is the founding act of a new landscape. Any incorporation into a landscape already shaped by Man allows that landscape to evolve so as to embrace a new quality, which can be negative or positive.

Any direct relationship with nature, in the few cases where such is still accessible to the common person, occurs through the sublime. It is a purely emotional relationship, lacking semantic articulation. It is a hic et nunc with no evolutionary consequence, a pure aesthetic pleasure, once a privilege reserved for philosophers and contemplative travelers, but now apparently extended to the adventurous tourist. In our direct and unmediated contemplations of nature, we stand as completely detached observers, for whom, by definition, entry on the scene is precluded.

Our relationship with landscape is much more complex than that with nature, involving a conceptual relationship in which we can and should enter the scene, more as a director than as an actor. The landscape, urban or rural, in which we engage in manufacturing or leisure, in which we move and save resources, is humanity's stage; it is where we perform our actions and rituals, meet our needs, material and immaterial, and express our Weltanschauung.

Observers and observed

All of a sudden he exclaimed: "So that's what Schrödinger meant!" He was very excited. "Julie, remember the ending of Schrödinger objectivation article? 'Subject and object are only one. The barrier between them cannot be said to have broken down as a result of recent experience in the physical sciences, for this barrier does not exist.'" I puzzled over this statement ever since we discussed that article. Now, in view of what you said, I get it. In contemplation, one is in touch with the truth. It is the incorporation you talked about. One is in the Oneness of all, so to speak. There is no subject and no object. It is only when one's mind reverts to the discursive mode that one objectivizes, that there is a bifurcation into the duality of subject and object. Schrödinger must have experienced what he wrote, through contemplation. Everything is beginning to bound together.[1]

In the Twentieth Century, quantum physics rendered the obsolete idea that Man can be outside the observed system, regardless of how isolated one may be. The observer is part of the system, and the very act of observing produces physical changes to the system. This conclusion holds especially true with regard to the rela-

Rossignol全球总部，有机的、覆有木材的结构形状与建造场地周围的群山形状遥相呼应

Rossignol Global Headquarters, organic, timber-clad shape echoes the profile of the mountains that surround the site

关系方面。

然而，人类利用建筑这一重要工具来干预自然的最重要的方面是，通过标志的融合建立全新的语义系统。通过融合这些标志，自然变成了景观，为人类活动建立了场所。

人、建筑对象和场景之间的关系并不简单，事实上它们之间的界限相当模糊。在一个不止有三个维度且能够感知到的非欧几里德几何图形当中，主体和客体可以互换角色，观察者可以变为观察对象，观察对象也可以变为观察者。因此，建筑不仅是场景中的一个元素，还是其中的一个有利因素。

最后，还有一些建筑并没有标志出它们从观察者转变为观察对象，在这些建筑中拥有一种视觉和空间上的连续感。现代建筑技术已经能够使内外之间的界限消失，但即使是在大量使用玻璃和钢铁的时代以前，使用石头和木材打造出来的模糊空间（介于公共门廊的内外空间之间）在概念上也得到了相似的结果。提到所有古老空间模糊性的缩影，人们很容易就会想起具有帕拉迪奥建筑风格的奥林匹克剧院的城市舞台。

正如上文所提到的，现代性为这个方向开启了全新的可能性，不仅借助于全新的建筑技术，而且（甚至主要地）通过废除空间形式化方面的标准来完成。

在上文描述的这些理念的极端情况中，这些项目展现了文脉"内""外"融合的广泛可能性。在所有情况中，内外之间、观察者和观察对象之间的模糊性都有被文脉的戏剧性和独特性埋没的风险。只有最高品质的项目以及对地方特色有着深刻理解的建筑师，才能通过运用不同的形式和方法，维护一种同等的关系，并真正实现这些元素之间的融合。

作为地质外形塑造而融入

Rossignol全球总部/法国，Saint Jean de Moirans，La Buisse/Hérault Arnod建筑师事务所

阿尔卑斯山的景观一直为居住在山上的人们创造了一种特殊的直接接触。紧凑、巍峨的山脉与恶劣苛刻的环境共同形成了一种建筑灵感，将这座建筑以传统的方式融入景观之中。

然而，恶劣的天气条件需要将内部空间与周围的环境清晰地分离开来，当代的新技术却可以使山上居民沉浸在其对景观的深厚情感关系之

tionship between Man and the stage he occupies.

The most important aspect, however, of human interventions with nature via the critical tools of architecture, is the establishment of a semantic system through the incorporation of signs. By incorporating these signs, nature becomes landscape and sets the scene for anthropic action.

The relationship among Man, the objects and the stage, is not simple; indeed, it is rather ambiguous. Subjects and objects change roles. Observers become the observed, the observed become the observers, in a perceived geometry that is no longer Euclidean but that assumes far more than three dimensions. Thus, a building is not only an element of the scene, but a vantage point of it.

Finally, there are buildings in which the transition from observer to observed is not marked, in which there is a visual as well as a spatial continuity. Modern building techniques have made possible the loss of boundaries between inside and outside, but even before the era of the massive use of glass and steel, conceptually similar results were obtained via architectural gestures in stone and wood in the ambiguous spaces between the interior and exterior spaces of public porches. It is difficult, then, not to be reminded, as the epitome of all the ancient spatial ambiguities, of the city-stage of the Palladian Olympic Theatre.

Modernity opens up new possibilities in this direction, as mentioned above, not only by means of new building techniques, but also, and even primarily, via the collapse of canons regarding the formalization of spaces.

The projects examined herein represent a broad spectrum of the possibilities for the incorporation "in" and "off" the context, in situations that represent the extremes of the concepts set out above. The ambiguity between inside and outside, between observer and observed, in all cases risks being overwhelmed by the dramatic uniqueness of the context. Only the projects with the highest quality and the architects with a deep understanding of the genius loci can allow, in various forms and approaches, the maintenance of a relationship of parity, a genuine incorporation among the elements.

Incorporation as geological shaping

Rossignol Global Headquarters / La Buisse, Saint Jean de Moirans, France / Hérault Arnod Architectes

Alpine landscapes have always created a special direct contact for mountain-dwelling people. The close, imposing mountains together with the hard and demanding environment tend to inspire an architecture which is incorporated traditionally into the landscape.

However, where harsh weather requires a clear separation of internal spaces from the surrounding environment, new technologies now allow mountain-dwellers to indulge their deep and emotional relationship with the landscape. Rossignol is not just an ordinary

Kilden表演艺术中心，弯曲的橡木墙壁的设计灵感来源于海浪
Kilden Performing Arts Centre, curving oak walls are inspired by ocean waves

弯曲的屋顶是轻井泽别墅A中最特别的元素，这样设计的目的是为了像鸟翼一样遮挡住上层建筑
a bent roof, the most particular element in Villa A in Karuizawa, was intended to shelter the upper building like a bird's wing

中。Rossignol不仅仅是一家普通的制造公司，它的核心业务与山脉以及山区人民紧密相连。

正是出于这个原因，Rossignol公司建造的建筑通过大量使用薄框架、大面积的水晶板、反光或透明（取决于倾斜度）的高天窗，可以让那些为公司工作的人们尽可能少地与赋予人灵感的环境相分离。从概念上讲，建筑的顶部不仅仅是一个屋顶。从地质角度而不是仅从建筑角度来说，它是一个构造的对象，通过复杂的几何形状来模仿当地的地形，并且回避了清晰可辨的排水沟槽。

在这座建筑的规划方案中，巨大的流动式开放空间与办公室那较为呆板和传统的空间完美共存。在开放空间的内部设计中，人们可以辨别出一切事物（例如材料和形状）是如何描述阿尔卑斯山的景观的：木材的重要用途、形如岩石峭壁的长条弯曲墙面、具有缆车形状的电梯。

整座建筑的规划没有任何限制，无论是在形式上还是在功能上，它强化了融入景观的理念，使其在未来的扩建方向上没有任何倾向，这一结果部分是由于其在功能、设计、管理和生产方面没有任何层次结构而产生的。

作为戏剧场景融入

Kilden表演艺术中心/挪威克里斯蒂安桑/ALA建筑师事务所

当这座表演艺术中心竣工时，它就像一位仪表堂堂的杰出演员，出现在克里斯蒂安桑港湾的舞台上。尽管它那绝妙的外形被特意定位在这座壮观的建筑中有争议的一侧，它还是经过了非常谨慎的设计，以避免仅仅成为一座奢侈的建筑。该中心仅仅能把自己作为最新的建筑而不是最后一座枯燥无味的建筑地标穿插在过去十年的城市景观中。

与Rossignol的结构一样，覆盖在Kilden表演艺术中心上面的结构也不单单是一个屋顶，而是一个巨大的舞台。就像一组俄罗斯套娃一样，这座容纳有表演大厅的建筑也是一个独立的剧院。

这种外部方正的木质结构有着蜿蜒的内弧面，弧面与大面积的水晶板墙体相交，形成了一条波浪线，从而让人联想到海洋的样子、海岸线的天际线以及周围的岛屿。屋顶与玻璃立面的交汇处并不是那么清晰，模糊地介于海港的现实和幻想的虚拟区域之间。

在另一边，建筑结构那尖锐的、棱柱状的外部边缘与海港邻近仓库展开了丰富的对话，减轻了其奢侈和古怪方面的意图。在这种情况下，很难将该项目定义为一种融入环境的简单操作。顺便提一下，最初宣布的

manufacturing company. Its core business is intimately related to mountains and mountain people.

For that reason Rossignol's building allows those working for the company to enjoy a minimal separation from the inspiring environment through the extended use of thin frames and large crystal panes, reflective or transparent depending on the inclination of the sharp, high skylight. Conceptually speaking, the top of the building is more than just a roof. It is a tectonic object, in more of a geological than an architectural sense, mimicking the local orography via its complex geometry, shunning a recognizable gutter line.

In the building's plan, large and flowing open spaces coexist with the more rigid and conventional spaces of the offices. In the interior design of the open space, one may discern how everything, materials and shapes, speaks of the alpine landscape: a significant use of wood, the long flexuous wall shaped like a rock cliff, the elevator in the shape of a cable car.

The plan of the entire structure acknowledges no limits, formally or functionally, reinforcing the concept of incorporation into the landscape and allowing for future extensions with no preference as to direction, thanks in part to the absence of any hierarchy among function, design, administration and production.

Incorporation as theatrical scene

Kilden Performing Arts Centre / Kristiansand, Norway / ALA Architects

Finalized for the performing arts centre, this building presents itself as an imposing, brilliant actor on the stage of the harbor bay of Kristiansand. Although its stunning shape is intentionally positioned on the controversial side of spectacular architecture, the building design is very careful to avoid being merely an extravagant object, able only to establish itself as the latest but not the last of the series of quite boring architectural landmarks that have punctuated urban landscapes in the last decade.

Like the Rossignol structure, the Kilden Performing Arts Centre is capped by a structure that is not a simple roof, but more of a huge proscenium. Like a set of Russian dolls, the building housing the performing halls is a theatre in its own right.

The intersection of the sinuous intrados of this wooden, externally squared structure with the large wall of crystal panes generates a wavy line evoking a marine pattern as well as the skyline of the shoreline and the surrounding islands. Where the roof meets the glass facade plays an ambiguity between the reality of the harbor and the fictitious domain of the fantasy.

On the other side, the sharp, prismatic exterior edge of the structure is in fruitful dialogue with the neighboring warehouses of the harbor, mitigating any intention of extravagance and eccentricity. It is difficult, in this case, to define this project as a straightforward operation of incorporation into the environment. By the way, the originally declared program of shaping the building "to act as

照片提供：Atelier Fournier-Maccagnan (©Thomas Jantscher)

将拉沃葡萄酒品酒中心有效地融合在景观之内，使其宛如一块岩石
effectively integrated Wine Tasting Centre in Lavaux into the landscape, similar to a block of rock

项目是将该建筑打造为“城市景观中的标志”，从而形成强大的张力和戏剧性，这在某种程度上使建筑与周围环境搭配起来，还可以让人回想起贝多芬关于第六交响曲的注解：“Mehr Ausdruck der Empfindung als Malerei（比如画的景色更多的是情感的表达）”。

通过透明度来融入

轻井泽别墅A/日本长野县/Satoshi Okada建筑师事务所

对于那些在日本文化中形成的建筑，融入景观的理念是一项更容易、更传统的任务。毫不意外的是，我们在轻井泽地区发现了建筑与建造场地关系的相似实例，它们更加注重自然方面的考量而不是景观。

若要放弃一个词汇，这个词绝对不会是“自然”，而是“强大”或“人为”，Satoshi Okada建筑师事务所已经为这座住宅设计了狭窄、翼状的表皮，它像武士刀中的ô-kissaki一样锋利、尖锐。那种表皮呈弯曲状，以确保邻近别墅（周围环境中唯一一座人造建筑）不会窥探到该住宅的隐私。

极简的生活空间完全与树林融合。弯曲的屋顶没有遮挡住这儿的景色，该建筑与环境的连续性是绝对的，整个内部空间都可以观赏到葱翠、高大的树木，连续不断的玻璃墙会挡住很小一部分的视线。室内生活空间的存在似乎并没有改变对周围环境的感知，这种感知代表了一个通过透明度来融入到景观之中的极端实例。

相比之下，容纳私人空间的半地下楼层则是石质的，非常坚固，通过优美地遵循着地形的轮廓，从而促进了与环境的融合。

作为模仿物融入

拉沃葡萄酒品酒中心/瑞士/d′ architects Fournier-Maccagnan工作室

Pascal Fournier和Sandra Maccagnant在该项目上采用了一种策略，他们将该建筑与引人注目的景观相融合，该景观一直向下延伸至莱蒙湖——一处人造景观，梯田、石墙、陡峭的斜坡以及葡萄园中那些规则的绿色直线占据了该景观的主要部分。

在这里，人造物完全融合到了景观中。尽管建筑呈现出直角边缘，不过Daniel Schleapfer设计的具有滤镜效果的钢丝网展示出了葡萄藤上叶子的光域，这使得人们可以想象建筑物的存在。建筑物棱柱状的体量通过类比的方法与周围的岩石和石墙融为一体。无钢混凝土和当地的石材模拟出上面阶梯状的葡萄园中那壮观的墙壁。小型葡萄园顶部的石板完成了建筑与景观融合的最后一步。

a sign in the cityscape” thus creates strong tension and drama, collocating the building with the environment in a way which recalls Beethoven’s note concerning his Sixth Symphony: “Mehr Ausdruck der Empfindung als Malerei (More Expression of Feeling than Picturesque Scenery)”.

Incorporation by transparency

Villa A in Karuizawa / Nagano, Japan / Satoshi Okada Architects

The concept of incorporation into the landscape is an easier, more traditional task for those operating within Japanese culture, and it is no accident that in the Karuizawa area we find similar examples of a relationship with the site that still comes very close to a consideration of nature rather than of landscape.

With renouncing a vocabulary which is in no way “naturalistic”, but is strong and anthropic, Satoshi Okada Architects have conceived the house under a narrow, winged surface, as sharp and pointed as the ô-kissaki of a katana sword. That surface is bent to protect the house from the view of a neighboring villa, the only artificial object in the surroundings.

The minimalistic living space is totally merged with the woods. Where the view is not shielded by the bent roof, the continuity with the environment is absolute, to the point of allowing a view of the beautiful tall trees throughout the interior space, lightly limited by a continuous glass wall. The perception of the surrounding environment seems unaltered by the presence of the interior living space, representing an extreme example of incorporation through transparency.

By contrast, the semi-underground level that houses the private rooms is stony and solid, contributing to the environmental integration by beautifully following the contours of the terrain.

Incorporation as mimesis

Wine Tasting Centre in Lavaux / Switzerland / Atelier d’architects Fournier-Maccagnan

Pascal Fournier and Sandra Maccagnant adopt a strategy in this project of merging the building into a dramatic landscape that slides down into Lake Léman – a landscape forged by Man, dominated by terraces and stone walls, steep slopes, and the regular, green lines of the vineyards.

Here the incorporation of the human-made is absolute. Despite the structure’s squared edges, nothing but the pixelated steel mesh by Daniel Schleapfer representing a raster of vine leaves allows one to imagine the presence of the building. Its prismatic volumes are melted analogically with the surrounding rocks and stone walls. Plain concrete and local stone mimic the imposing walls of the terraced vineyards above. On the top, the slabs make small vineyards complete the process of incorporation into the landscape.

In accordance with the functional requirements of a wine cellar, the internal spaces of this tasting room are related only to the

照片提供：Circa Morris-Nunn Architects (©Peter Whyte)

Saffire度假村，特意再现了海岸形状
Saffire Resort, intentionally evocative of coastal forms

照片提供：Reiulf Ramstad Architects

Troll Wall餐厅映现了动态的山脉景观
Troll Wall Restaurant reflecting the dynamic mountain-scape

根据酒窖的功能要求，品酒室的内部空间只与土地的中心部分有关联，并打造出了一种朦胧的地下景观，该景观只能通过Schleapfer设计的嵌板上的密目网来与外界接触。

通过类比方法融入

Troll Wall餐厅/挪威，Møre og Romsdal，Trollveggen/Reiulf Ramstad建筑师事务所

该建筑位于欧洲最高的垂直岩壁的山脚下，面对着北欧最壮观的山岳景观之一。建筑师们已然接受了目前的趋势——让他们的项目成为当地的地标和建筑景点，利用带有明显折角的线条和垂直线条进行设计，这唤起了人们对于大教堂的尖顶部分的记忆，同时也让人们联想到了景观的垂直状态。

餐厅棱角分明的体量交替使用了玻璃和混凝土。平面玻璃墙展现出了一种非正交结构，这有助于将山墙的巨大纹理与镜像景观和其内部的透明感知融合在一起。

已规划好的项目通过极其有限的材料和形状方面的语汇得以优美实现。建筑师对技术细节的考虑细致入微，这也使得建筑物精致、简洁，消除了任何多余的元素，以防止观察者的注意力从建筑物与周围群山的类比中分散。

作为内在状态融入

Saffire度假村/澳大利亚科尔斯湾/Circa Morris-Nunn建筑师事务所

尽管如建筑师所说，这家度假村再现了沿海地形，但同时它那波浪状的弯曲表面也反映出了旁观者的一种内在状态，而非简简单单地模拟建造场地的形状。

根据在观测点所观察到的，度假村主建筑的屋顶能够使人们联想到海湾柔和的天际线或一种海洋生物的形状，同时也能够唤醒人们对令人安心的原始宁静的记忆。显而易见的是，周围荒野的极度纯净会促进人们对一个地方特色的认知过程，但还是要采取一切预防措施来消除可能与该地点相对立的元素。

在从原始自然演变到控制良好、可持续的人文景观这一艰巨过程中，套房零散弯曲的外观的低姿态与主建筑融合在一起。

hearth of the earth, generating a shadowy underground landscape that comes into contact with the outside only through the dense screen of Schleapfer's panel.

Incorporation by analogy

Troll Wall Restaurant / Trollveggen, Møre og Romsdal, Norway / Reiulf Ramstad Architects

Sitting at the foot of the tallest vertical rock face in Europe, this building confronts one of the most spectacular mountain landscapes in Northern Europe. The architects have accepted the current trend of letting their project play the role of local landmark and architectural attraction, designing strongly angled and vertical lines, evoking a cathedral spire that recalls the landscape's verticality.

The edgy volumes of the restaurant alternate glass and concrete. The plain glass walls show a non-orthogonal structure that helps to merge the macrotexture of the mountain walls with both the mirrored landscape and its transparent perception from the interior.

The programmed task is beautifully achieved via an extremely limited vocabulary of materials and shapes. Exquisite attention to technological detail enforces the building's sophisticated simplicity, removing any redundant elements that might distract the observer from the analogy of the architectural object with the surrounding mountains.

Incorporation as inner state

Saffire Resort / Coles Bay, Australia / Circa Morris-Nunn Architects

Although, as stated by the architects, the shape of the resort is evocative of coastal forms, at the same time its wavy, sinuous surface is the reflection of an inner state of the beholder and not simply the expression of a mimetic relationship to the site.

Depending on the observation point, the roof of the resort main building may remind the soft skyline of the bay or the shape of a marine creature, awakening memories of reassuring ancestral peacefulness. Although it is evident that the process of identification with the genius loci is facilitated by the exceptional purity of the surrounding wilderness, every precaution was taken to eliminate any element that might conflict with the place.

The low profile of the fragmented curve of the suites joins the main building in the difficult task of evolving the pristine nature in a well controlled and sustainable human landscape. Aldo Vanini

1 《大自然喜欢藏起来：量子物理和现实》，西方的视角，Malin, S.，牛津大学出版社，2001年，第123页。

1. *Nature Loves to Hide: Quantum Physics and Reality*, a Western Perspective, Malin, S., Oxford University Press, 2001, p.123

Saffire度假村

Circa Morris-Nunn Architects

该场地位于澳大利亚塔斯马尼亚岛东海岸的科尔斯湾，俯瞰着大牡蛎湾、Hazards海滩以及菲欣纳半岛。建筑师的提案对拥有20间私人套房的私密豪华度假村进行了设计，旨在使其凭借自身的力量成为一个旅游胜地，并且主要为来此地的（州际或国际）客人提供服务。它将在该国提供（并且可能创造）一个新的高端旅游市场。对甲方而言同样重要的是，该项目具有一个标志性的、非常容易识别的形状，同时该形状也与自然场地相关联。虽然度假村位于广阔的天然原生海滨景观之中，但该项目场地却因其以前被当作一个废弃的旅行车停车场而伤痕累累。该项目不仅修复了这片场地，对其独特的品质进行了诠释，还创造了一个可以进行体验的空间。该度假村与其场地形成了一种有机的关系，并且特意唤起人们对海岸形状的记忆。

然而，该形状也与该项目提案的需求相呼应。事实上，度假村主建筑（或"保护区"）的入口设置在建筑尾部，其室内朝向一个大型体量开放，置身该体量中可以将Hazards海滩与大牡蛎湾的全景一览无余。接待处休息室位于这一层，中间楼层设有酒吧、餐厅和休息区，最底层设有画廊、会议室及日间水疗室。私密的通道将接待楼和套房连接起来，套房间隔排列以展现其间的景观。Hazards海滩是该地的主要特色，该建筑有意地遮挡和展现了贯穿整个旅程的风景。

该设计的主要挑战之一是调和甲方和市场对奢华的认知，同时还要捕捉到独特的塔斯马尼亚文脉元素。风景是其重要的组成部分，然而，材料、颜色、规模和形式也起到了决定性作用。材料的选用经过了多方面的考量——实际原因（如建筑形式、实用性和远程施工）、它们的美学关联（地方风格、自然文脉，或对豪华的期望），以及减少空间或体积。这通过一块镀灰板得到了加强，该板是建筑师对周围景观仔细审视之后所进行的尝试。此外，度假村宏伟的主建筑与周围更大的背景环境相呼应，同时又与更具个性化和小型规模的套房相中和。

凡被认为可能出现的有关可持续性设计的问题都经过了深思熟虑，然而，这也平衡了关于偏远选址和豪华度假村的市场期望（无论是在材料的选择上，还是在建筑设备方面）等施工问题。该项目需要考虑的主要问题是在这样一个典型的受干旱影响的地区的用水问题。作为该项目一部分的新雨水收集与存储设施以及污水处理设施被建在了场外，可供该开发项目和附近的乡镇使用。屋顶的雨水也被收集和再利用于倒影池以及特定的节水装置。由于该地区以凉爽气候为主，且场地朝向南面，因此另一个重要因素就是该度假村的供暖问题。所有建筑都经过了良好的保温处理，且安装了高性能的玻璃。该项目还采用了节能的水暖和空调系统。

Saffire Resort

The site is located at Coles Bay on the east coast of Tasmania, Australia and overlooks Great Oyster Bay, the Hazards and the Freycinet Peninsula. Architects' brief was for an intimate luxury resort of 20 private suites that was intended to be a destination in its own right and mainly caters to inbound – interstate or international –

guests. It will provide for (and possibly create) a new high-end tourist market in the state. Also important to the client was that the project had an iconic, highly recognizable form that also related to the natural site. Although the resort is located within an extensively natural native coastal landscape, the project site was scarred from its previous use as a disused caravan park. The project became as much about repairing the site and interpreting its unique qualities as it was about creating a space from which it could be experienced. The resort is organic in its relationship to the site and is intentionally evocative of coastal forms.

However, the form also works with the programmatic requirements of the brief. Essentially the resort main building (or sanctuary) is entered from the tail and its inside opens into a large volume that provides a panoramic view of the Hazards and Great Oyster Bay. The reception lounge is on this level with the mid level containing bar, restaurant and lounge areas and lowest level gallery, board room and day spa. Undercover walkways link the reception building to the suites which are spaced to reveal views through. The Hazards are a dominate feature of the site and the architecture deliberately blocks and reveals the views throughout the whole journey.

One of the main challenges of the design was to reconcile the client and market's perception of luxury and still captures elements of the unique Tasmanian context. The view is important part of this, however, materials, colors, scale and form also contribute strongly. Materials were selected to work on many levels – pragmatic reasons (such as built form, availability and remote construction), their aesthetic associations (vernacular style, natural context, or luxury expectations) and to extenuate spaces or volume. This was reinforced by a pallet that was sampled from a close examination of the surrounding landscape. Furthermore, the grandness of the resort main building, which responds to the larger context, is counteracted by a more personal and intimate scale in the suites.

All the possible issues relating to sustainable design were considered, however this was also balanced against construction issues on a remote site and market expectations for luxury resorts (both in material selection and building services). A major consideration for the project was water usage in the place which is typically affected by drought. New rain water collection & storage infrastructure were built (off site) as part of the project for use by the development and the nearby township as well as sewage treatment facilities. Rainwater from roofs is also collected and reused in the reflection pools, as well as water efficient devices being specified. Another important factor, due to the predominately cool climate and south facing site, was heating the resort. All buildings are well insulated and high performance glazing was installed. Energy efficient water heating and air conditioning systems were used. Circa Morris-Nunn Architects

项目名称：Saffire Resort 地点：Coles Bay, Australia
建筑师：Circa Morris-Nunn
项目团队：Robert Morris-Nunn, Peter Walker, Poppy Taylor, Ganche Chua, Jarrod Hughes, Judi Davis, Chris Roberts, Kylee Scott, Gary Fleming, Tina Curtis
建造商：Fairbrother Construction
项目管理：Stanton Management Group
结构工程师：Gandy+Roberts 服务工程师：Wood+Grieves
立面工程师：Hyder Consulting
紧急停车系统：Wood+Grieve Engineers 照明：Point of View
景观设计：Inspiring Place
室内设计：Chada(in collaboration with Circa Morris-Nunn)
甲方：The Federal Group 用地面积：119,000m²
建筑面积：internal_3,640m², outdoor undercover_1,225m²
竣工时间：2011 造价：$ 32 million
摄影师：©Peter Whyte(courtesy of the architect)-p.30^{bottom}, p.32~33, p.33^{bottom}
©George Apostolidis(courtesy of the architect)-p.26~27, p.29, p.30^{top}, p.33^{middle}

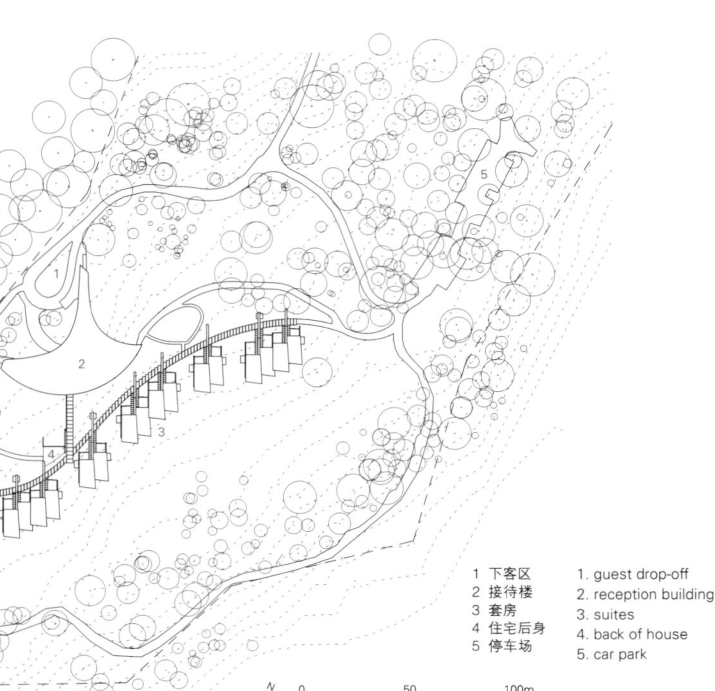

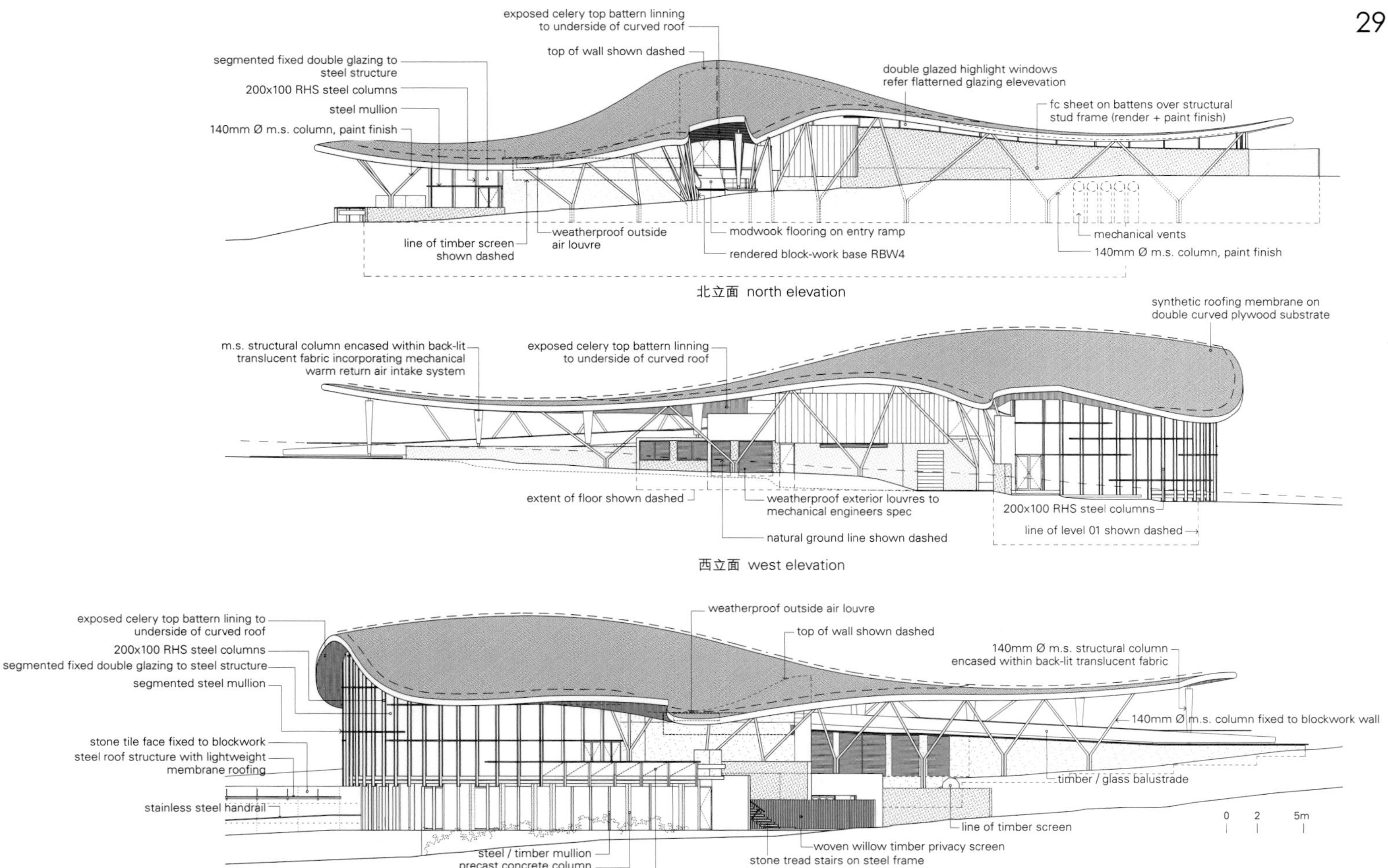

北立面 north elevation

西立面 west elevation

东立面 east elevation

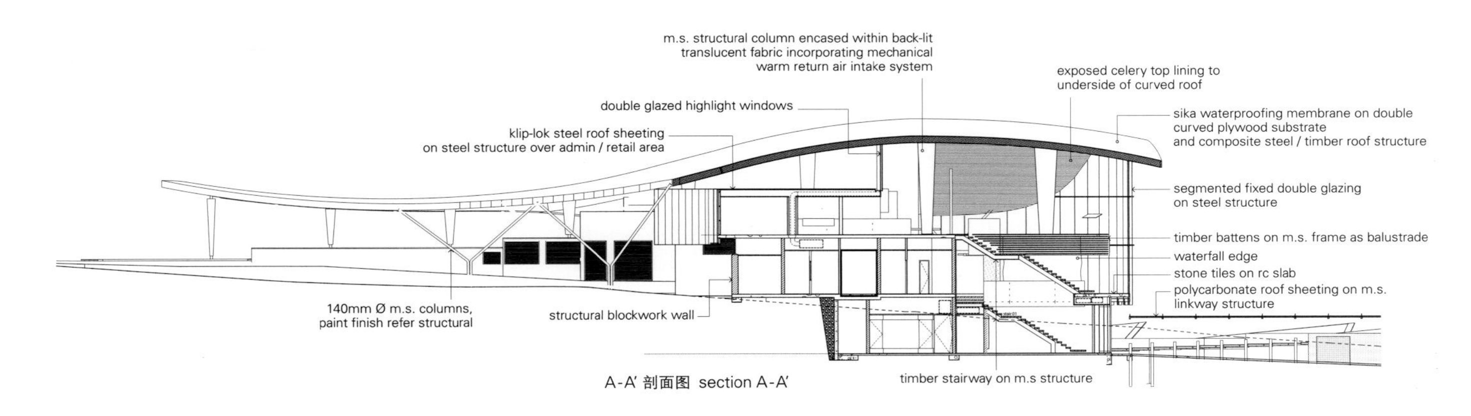

A-A' 剖面图 section A-A'

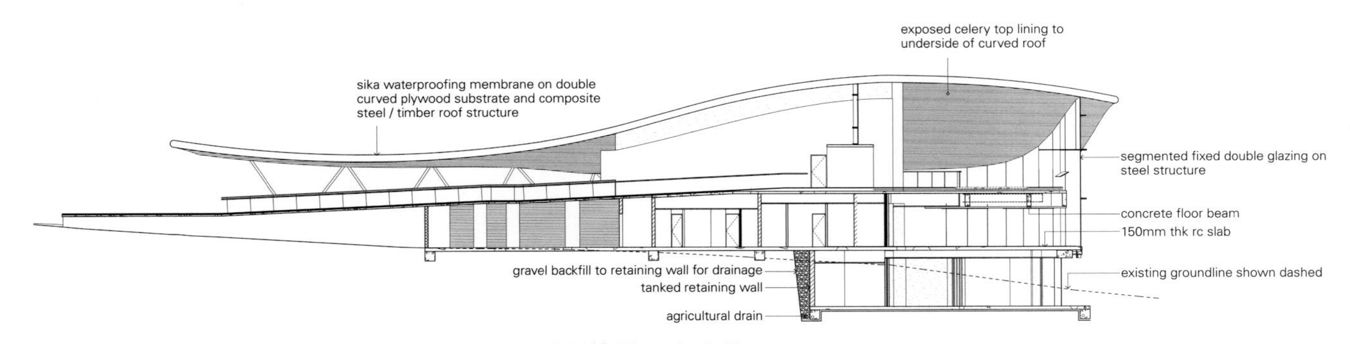

B-B' 剖面图 section B-B'

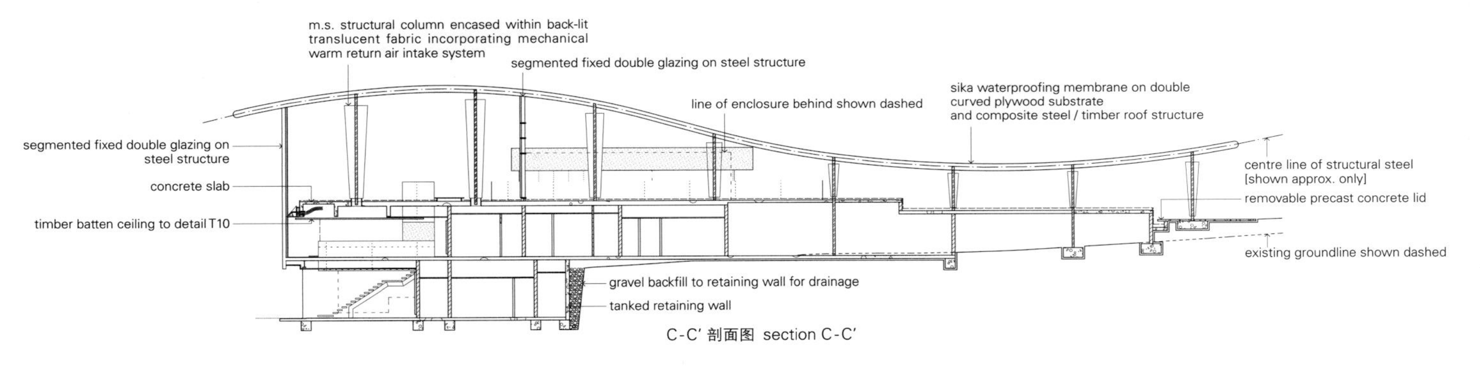

C-C' 剖面图 section C-C'

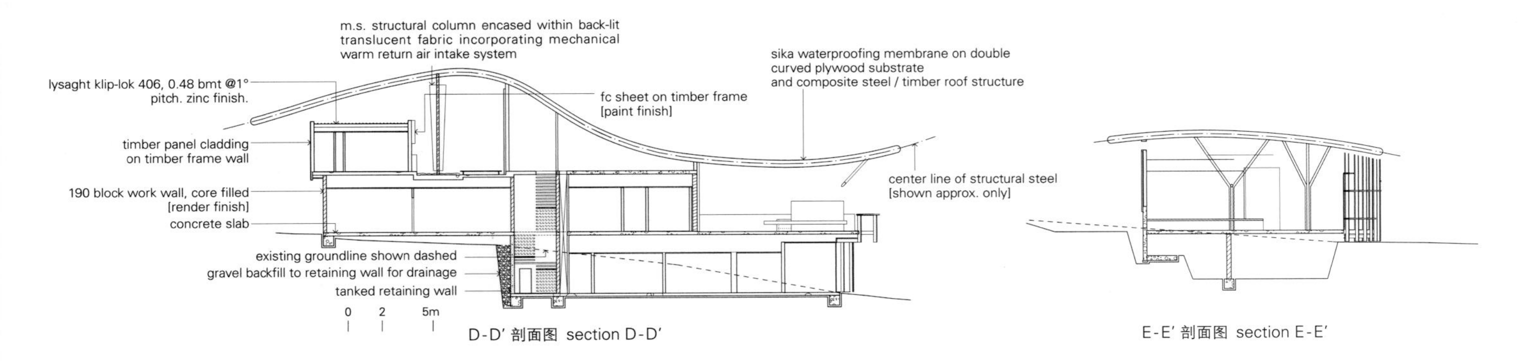

D-D' 剖面图 section D-D'

E-E' 剖面图 section E-E'

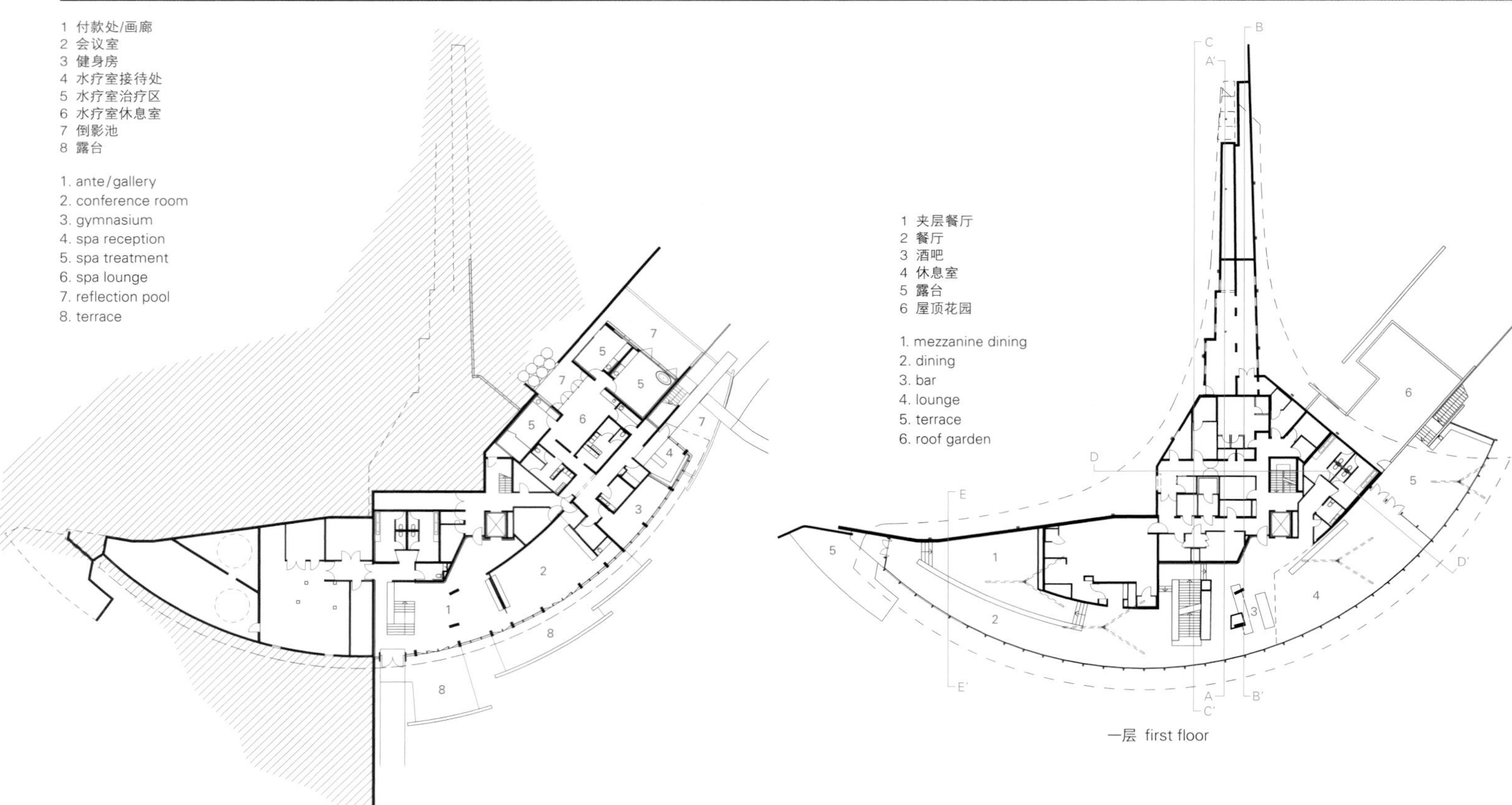

地下一层 first floor below ground

一层 first floor

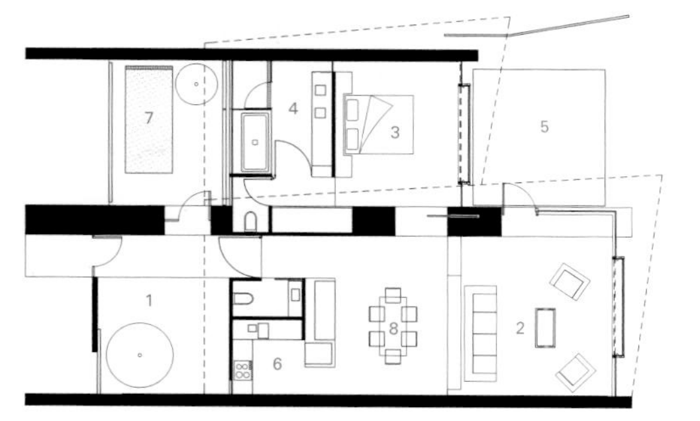

高级套房层 premium suite floor

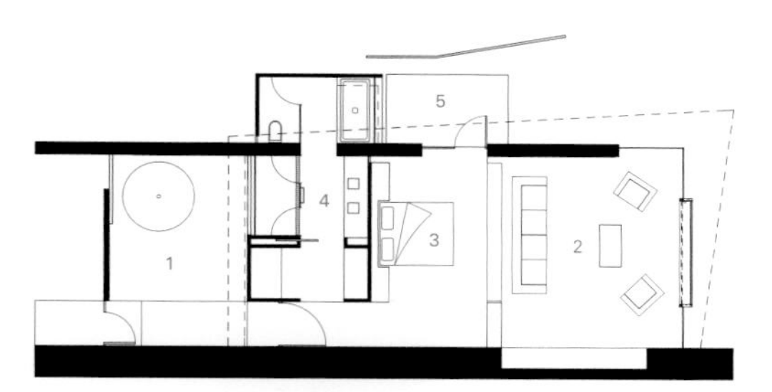

奢华套房层 luxury suite floor

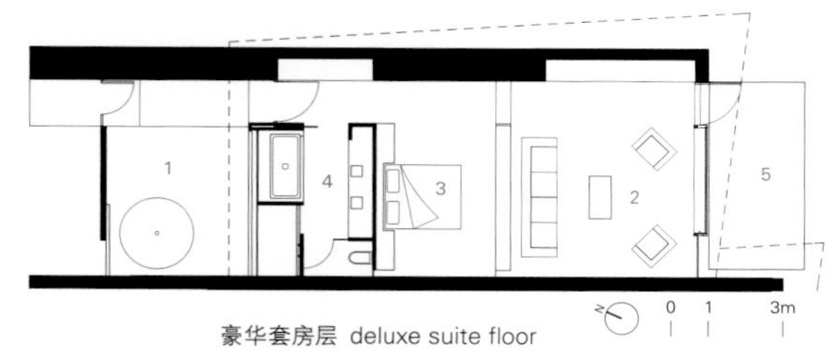

豪华套房层 deluxe suite floor

1 入口庭院 2 休息室 3 卧室 4 浴室 5 露天平台 6 厨房 7 按摩池 8 餐厅

1. entry courtyard 2. lounge 3. bedroom 4. bathroom 5. deck 6. kitchen 7. Jacuzzi pool 8. dinning

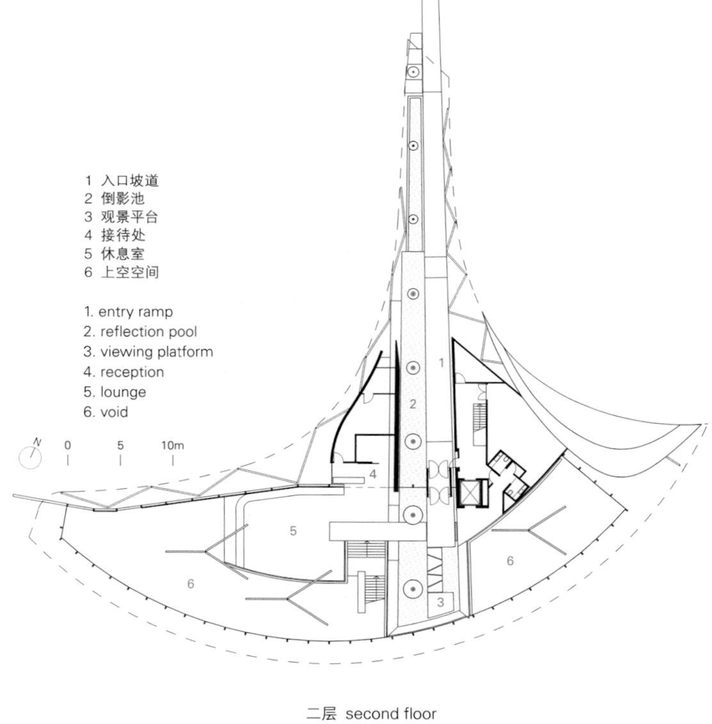

二层 second floor

Rossignol全球总部

Hérault Arnod Architectes

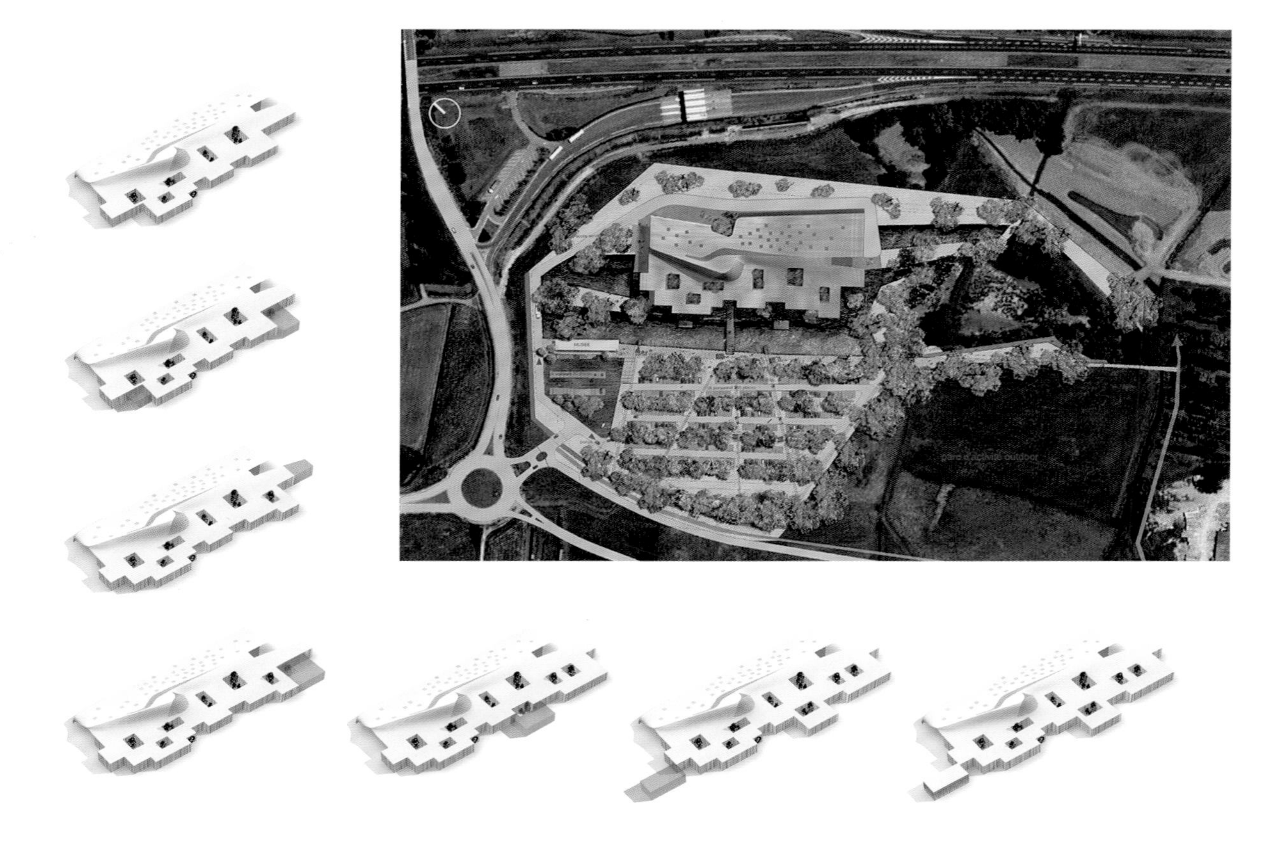

Rossignol是国际滑雪品牌的历史性领跑者，该品牌与山和雪有着直接联系。所以Rossignol全球总部项目没有被打造成一座老式的办公楼，而是在设计中借鉴了自然和山峰的特点，同时它也融合了技术，这与顶级体育运动是不可分割的。

场地位于被群山环绕的平原中心，这里是原有农田的延伸，潮湿且极为平整，北侧是里昂到格勒诺布尔的高速公路。这座建筑是专为Rossignol这一品牌而设计的，它是公司品牌实际功能和美好憧憬的结合，建筑外形令人惊叹并且极其简约，设计受到滑板运动的启发，受到流畅运动的启发，同样也受到浮雕以及自然力量所塑造的积雪和冰川的启发。屋顶覆盖了整个项目，并与景观和自然形成互动。该建筑由木材覆盖的有机形式与环绕场地的山体的轮廓相呼应。

屋顶覆盖了三种类型的空间：

— 滑雪比赛用品的生产车间、该品牌的技术展示处以及技术室，所有房间均沿着高速公路集中设置。

— 办公楼层，其中包括行政部门和销售部门，研发、研究和设计部门等。

— 壮观而又明亮的通道，从建筑的一侧穿到另一侧的社交会面空间。通道末端变宽成为陈列室。

在高速公路这边，公司标志的重复设置使立面呈现出动态效果。建筑正立面向上抬起，在车间上方形成屋顶，到达顶端后在西南侧向下延伸，以覆盖办公区域，然后与种有白桦树的庭院相交，这些树似乎穿过了屋顶，从而使自然与建筑交织在一起。屋顶和办公室立面不规则的外形为未来所需的扩展提供了机会。附加建筑可以在不破坏项目的平衡和特性的情况下建造。

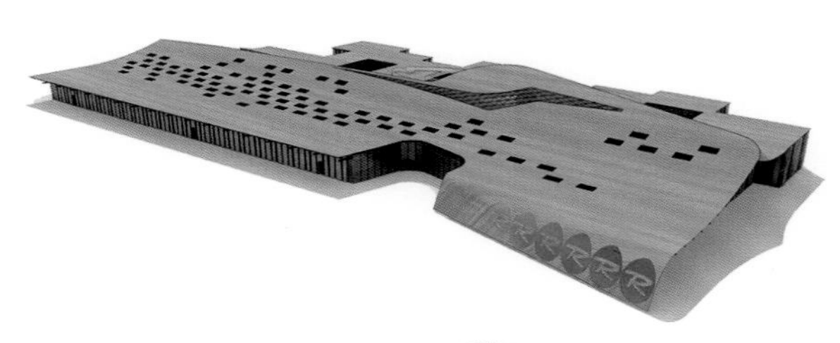

在里面，建筑像"蜂巢"一样发挥着作用，不同的功能既相互联系又相互作用，人们享受着一起工作和彼此会面的经历。为了鼓励内部沟通，社交空间分布在建筑的各处。餐厅位于建筑的顶部和通道的重力中心，被设计为公司活动的主要地点：两个大型的玻璃屋顶将天空和山脉的全景分隔开，一边面向Vercors，而另一边面向Chartreuse。

外部围护结构仅使用了两种材料：木材（天然落叶松）和玻璃。该结构由钢制成，像一个有机的骨架，用其多个弯曲的表面描绘出建筑的形状。人们在车间和办公室可看到屋顶框架。设备楼层的梁柱框架横跨12m~15m，该跨度提供了尽可能自由的空间。

该建筑采用了对环境影响最小的设计。所选用的技术产生了一座高效且节能的建筑，它具有良好的保温隔热性能，通过屋顶上的木材遮挡夏日的阳光。建筑系统得到了优化——车间机器产生的热量被回收并重新注入到供热网。办公室通过自动开启的窗户可进行自然通风。

Rossignol Global Headquarters

The image of Rossignol, a historic leader in the world of skiing, is intimately linked to the mountains and snow. The project for its global headquarters has nothing to do with the stereotypical office building, but is a tribute to nature and to the peaks, but also to technology, which is inseparable from top-level sport.

The plot stands in the middle of a plain surrounded by mountains. It is a stretch of former farmland, marshy and perfectly flat, bounded on the northern side by the Lyon-Grenoble motorway. The architecture has been designed specifically for Rossignol, a fusion of the company's functional and fantasy aspects, in a surprising and minimalist form: it is inspired by board sports, by fluidity of motion, and also by relief, snow and glaciers sculpted by the elements. The roof, which envelops the whole project, is topographically in osmosis with nature and the landscape. Its organic, timber-clad shape echoes the profile of the mountains that surround the site.

项目名称：Rossignol Global Headquarters
地点：Centr'alp 2 - La Buisse - Saint Jean de Moirans (38)
建筑师：Hérault Arnod Architectes
项目团队：Jérôme Moenne-Loccoz (project manager), Alexandre Pachiaudi, Camille Bérar, Nicolas Broussous, Matthias Jäger
竞赛团队：Florent Bellet, Adela Ciurea, Israel Lopez Vargas, Alexandre Pachiaudi, François Deslaugiers (for the panoramic lift)
结构工程师：Batiserf 经济学工程师：Nicolas 景观工程师：Forgue
流体工程师：Cap Paysages
甲方：Skis Rossignol SAS
用途：offices and open spaces for administrative departments, R&D, design, etc., racing skis workshop, showrooms
总楼面面积：11,600m²
设计时间：2006 施工时间：2007 — 2009
摄影师：courtesy of the architect-p.38~39
©André Morin (courtesy of the architect)-p.34~35, p.37, p.40, p.41, p.43, p.44, p.47, p.48~49
©Gilles Cabella(courtesy of the architect)-p.36, p.42
©Marie Clérin(courtesy of the architect)-p.45

The roof covers three types of space:
- The racing ski production workshop, the brand's technological showcase, and technical rooms, all grouped alongside the motorway.
- The office floors, which include the administrative and sales departments, R&D, research and design, etc.
- The street, spectacular and bright, the space of social encounter, which crosses the building from side to side. At its end, the street widens to become the showroom.

On the motorway side, the facade creates a kinetic and dynamic effect reinforced by the repetition of the logo. The front of the building rises to form a roof over the workshops and then on to the apex, and descends again on the south-western side to cover the office area. It is then intercut with patios planted with birch trees that seem to grow through the roof: nature and building intertwine. The irregular profile of the roof and office facades leaves the opportunity for future extensions as required. Additions can be built without disrupting the balance and identity of the project.

Inside, the building functions like a "hive" in which the different functions come into contact and interact, where people enjoy the experience of working together and meeting each other. To encourage this internal communication, social spaces are distributed around the building. The restaurant, situated right at the top and at the gravity centre of the street, is designed as the primary locus for the company life: two great glass-roofs divide up the panoramic views to the sky and the mountains, on one side to the Vercors and on the other to the Chartreuse.

Only two materials are used for the external envelope: wood (natural larch) and glass. The structure is made of steel, like an organic skeleton that outlines the shape, with its multiple warped surfaces. The roof frame is visible in the workshop and offices. The post and beam frame of the service floors spans of 12 to 15 meters to leave the space as free as possible.

The building is designed for minimal environmental impact. The technical choices make it an efficient and energy-saving building, well insulated and protected from the summer sun by the timber over-roof. The systems are optimized – the heat produced by the workshop machines is recovered and re-injected into the heating network. The offices receive natural ventilation through automatic window opening.

Hérault Arnod Architectes

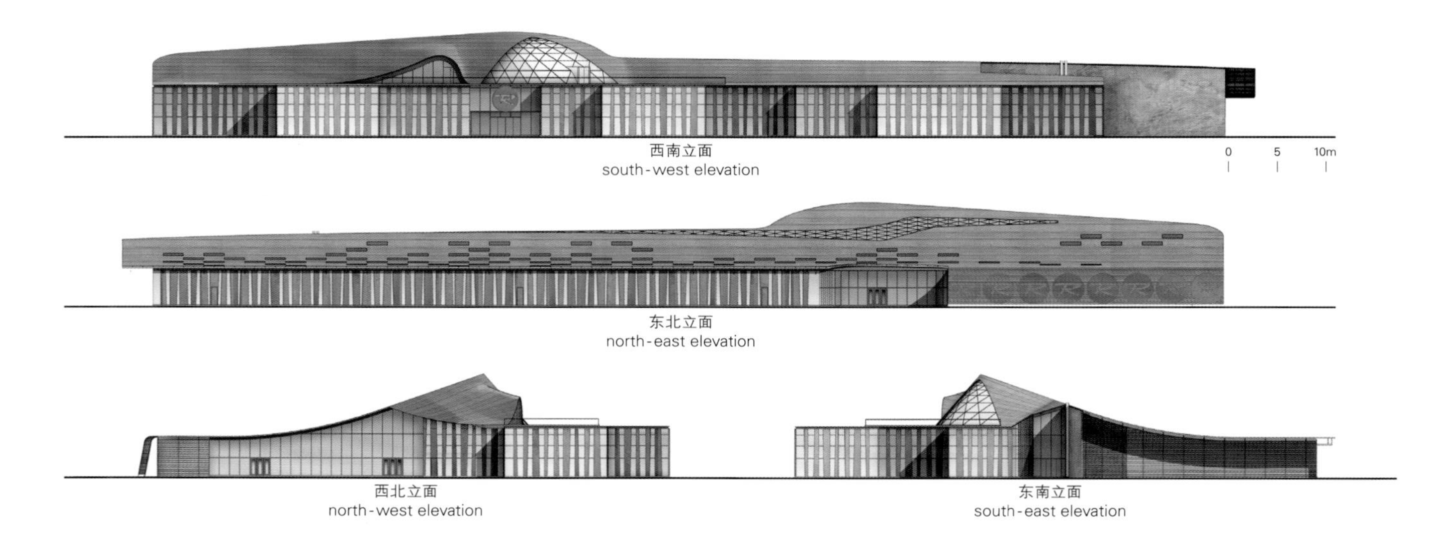
西南立面
south-west elevation
0
5
10m
东北立面
north-east elevation
西北立面
north-west elevation
东南立面
south-east elevation

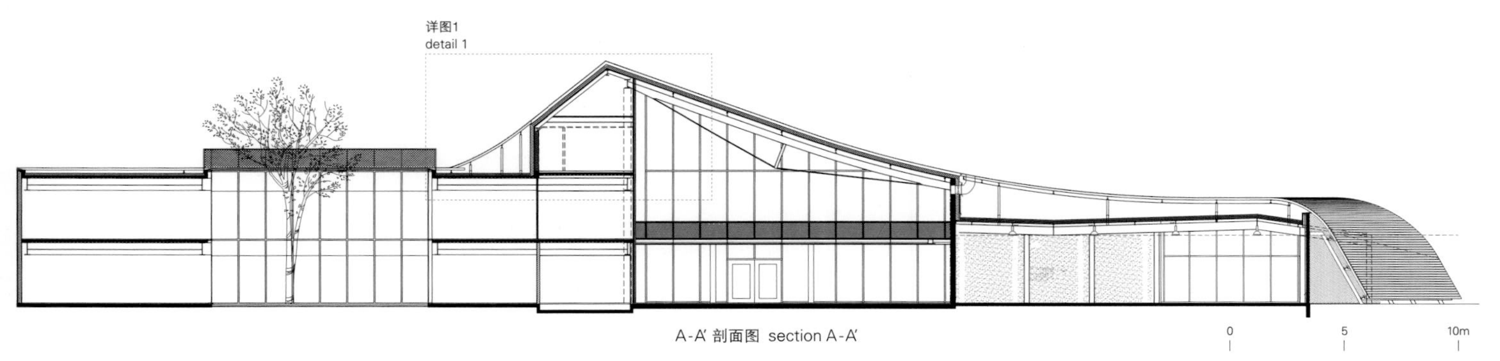

A-A' 剖面图 section A-A'

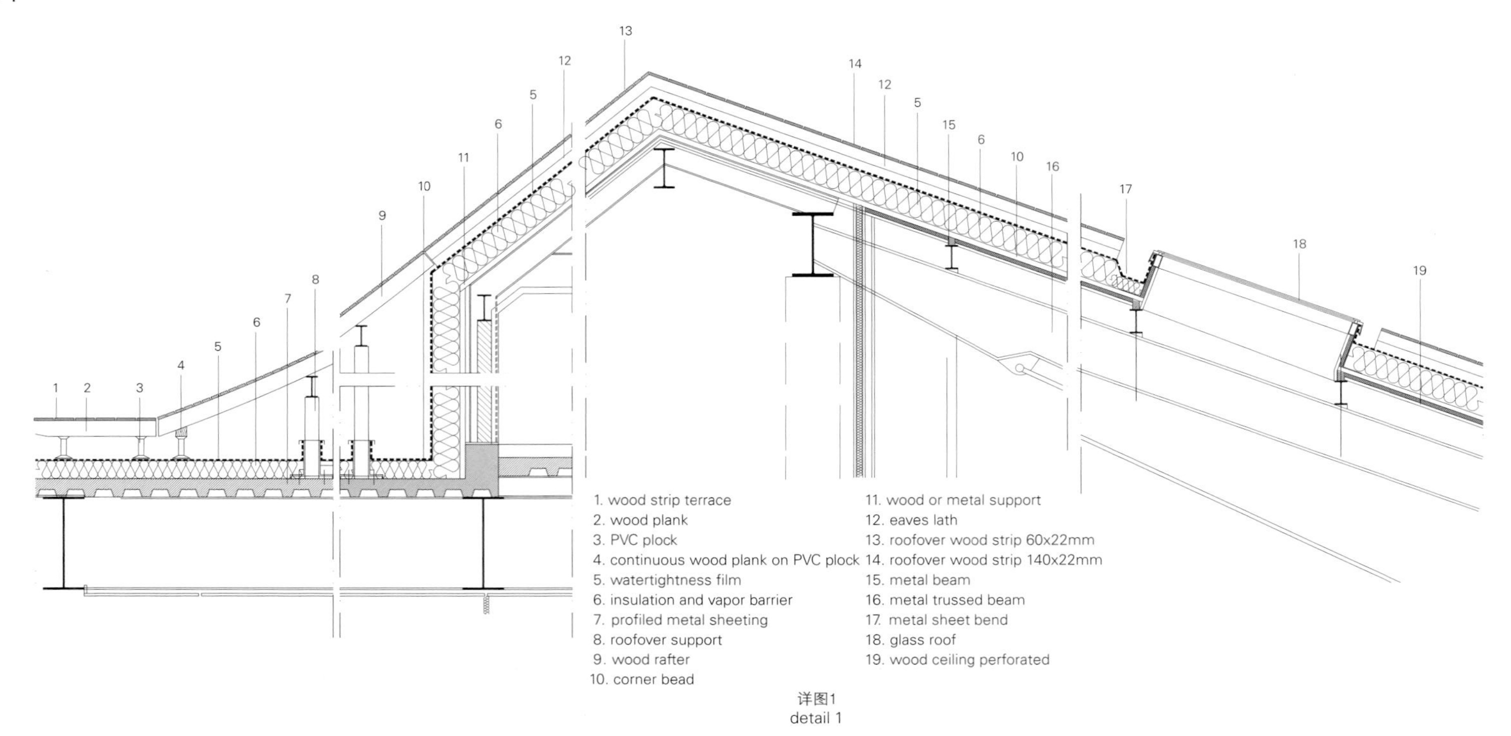
1. wood strip terrace
2. wood plank
3. PVC plock
4. continuous wood plank on PVC plock
5. watertightness film
6. insulation and vapor barrier
7. profiled metal sheeting
8. roofover support
9. wood rafter
10. corner bead
11. wood or metal support
12. eaves lath
13. roofover wood strip 60x22mm
14. roofover wood strip 140x22mm
15. metal beam
16. metal trussed beam
17. metal sheet bend
18. glass roof
19. wood ceiling perforated

详图1
detail 1

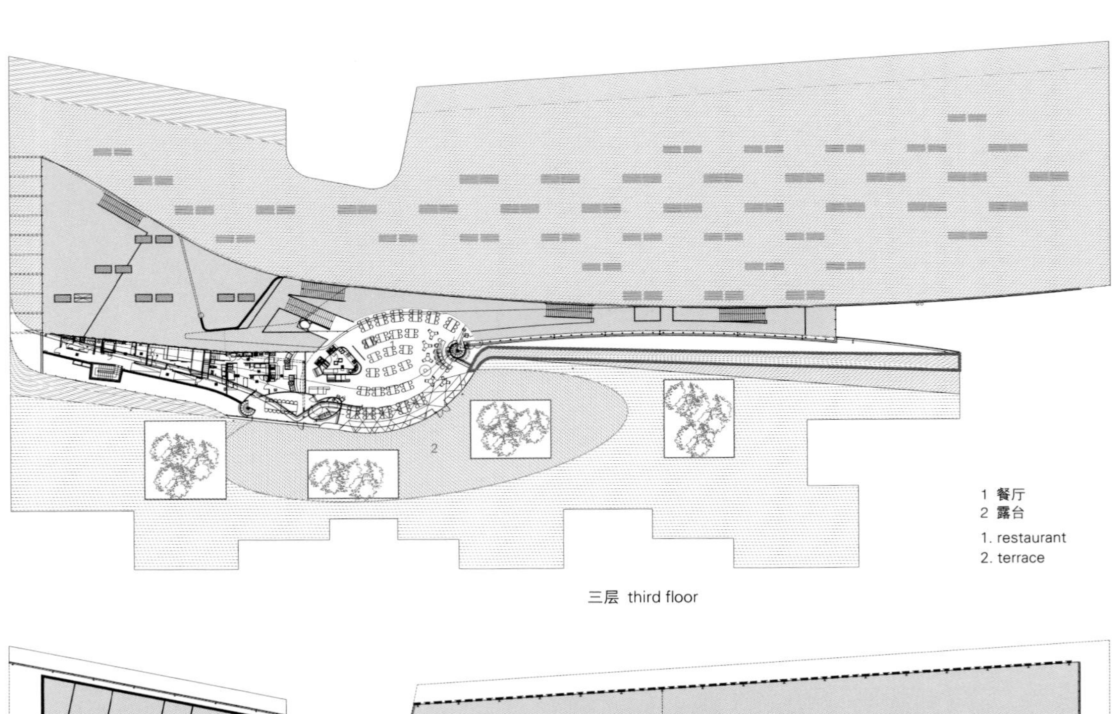

三层 third floor

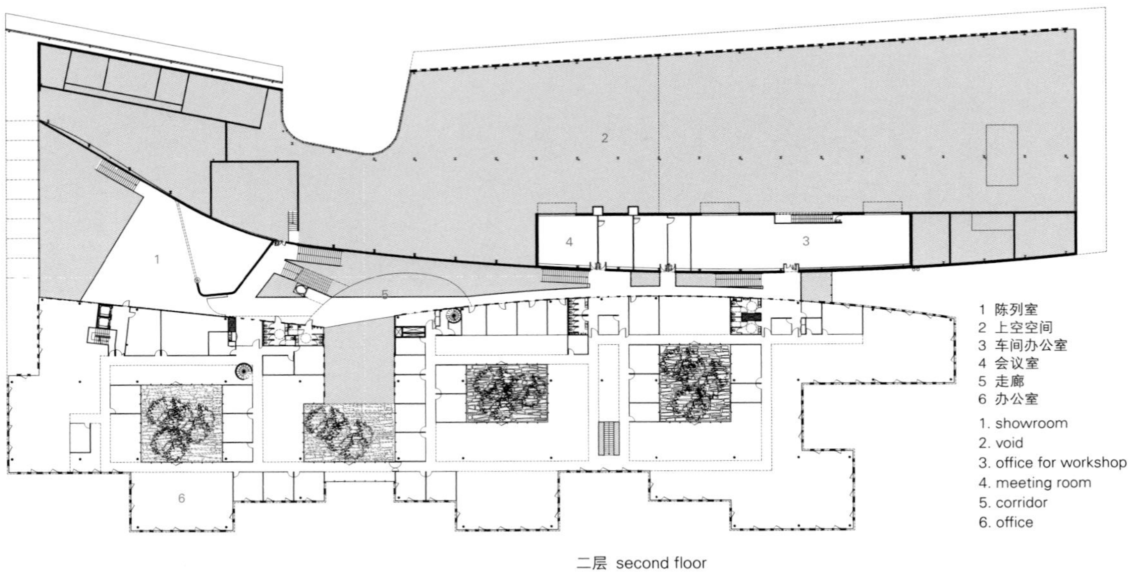

二层 second floor

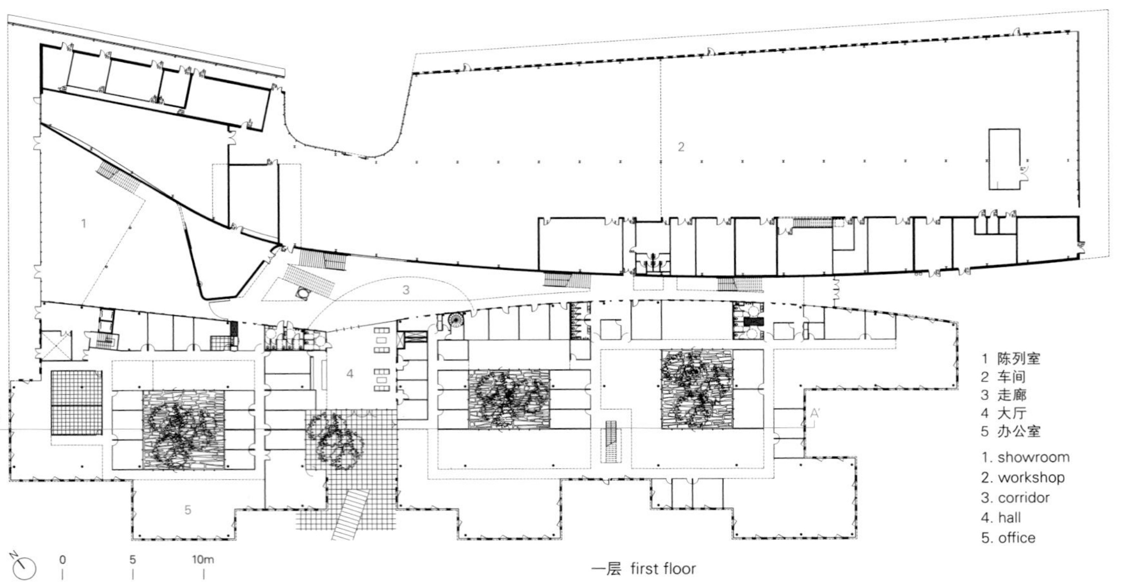

一层 first floor

拉沃葡萄酒品酒中心

Atelier d'architectes Fournier-Maccagnan

N
0
10
20m

项目名称：Wine Tasting Centre in Lavaux
地点：Route du Lac 2, CH-1071 Rivaz
建筑师：Atelier d'architectes Fournier-Maccagnan
视觉艺术家：Daniel Schleapfer
用地面积：2,535m²
建筑面积：330m²
总楼面面积：503m²
竣工时间：2010
摄影师：©Thomas Jantscher (courtesy of the architect)

该场地中的建筑致力于推广葡萄园和葡萄酒，该建筑所处的景观是受联合国教科文组织认定并保护的世界遗产，就该建筑的整合性问题，还需要更深一步的思考。该中心兼具独特和典型的特征，如梯田葡萄园以及源于Forestay的湖泊和溪流，既划分了土地，又体现出了一种独特的结构元素。

在梯田葡萄园的围合之下，建筑第一部分几乎是不可见的，但主体空间却冲破藩篱，以其独立的特性脱颖而出。就像在上面和后面都可以看见的村庄中的屋顶一样，建筑的外立面体现了手工制作的特点。混凝土由当地沙子和集料组成，在固定前经过了斜面切割，创造出了一种天然和手工相结合的效果，使之如Forestay大瀑布脚下的岩壁一样自然而有力。该建筑的文化韵味表现在由艺术家Daniel Schleapfer设计的以古怪风格展现葡萄藤和葡萄叶的悬置钢质画布上。

该建筑内部延伸至存放葡萄酒的空间。带有品酒吧台的接待处是这一理念的核心元素。宽敞的分类架将酒瓶沿着墙壁成行陈列，其暖色的橡木材质让人联想到酒桶，同时还可作为一个消音构件。你会被美酒包围，这里的灯光如地窖中的光线一样柔和，以便于保存酒。品酒的地点位于一个私密的空间内，可以促进人们进行自我反思。在这里的旅程中，游客们一路伴随着暗光，不时会遇到走廊和楼梯，唤起人们对典型的迷宫式葡萄酒地下仓库或狭窄的葡萄园小径的联想。这一探索之旅会将人们引至自助售酒机区域、地下室中的投影剧院以及上面楼层中的会议室。从这里，透过悬挂画布的网眼，可以将小瀑布、葡萄园和湖泊尽收眼底，这是该建筑中唯一一处可以看到这些风景的地方。

就地取材是该建筑理念的主要部分，材料包括：水泥和钢、葡萄园的特色构件，以及葡萄酒酿造过程中出现的橡木制品。

Wine Tasting Centre in Lavaux

The construction of a site dedicated to the promotion of vineyards and wine, located in a landscape recognized as a World Heritage by UNESCO and under its protection, requires some deep reflection on its integration. The site offers both unique and typical characteristics such as the terraced vineyard, the lake and the stream running from the Forestay, offering both a division of the land as well as unique structural elements.

The first body is barely discernible among the terraced vineyards that envelope it, while the main space breaks free and asserts itself with a separate identity. Like the roofs of the village visible above and behind, the facades reflect a manually worked character. Concrete composed of local sand and aggregate was incised with a rake before setting, creating an effect both raw and worked by a human hand, imposing itself forcefully and naturally like a rock face at the foot of the Forestay cataract. The cultural image of the building is exemplified by a hanging steel canvas representing grape vines and leaves in a pixilated stylization, created by the artist Daniel Schleapfer.

The interior spreads into spaces of wine storage. The reception with its wine-tasting bar is the central element of the concept. Spacious pigeonholes to hold and present the bottles line the walls, their warm oak reminding us of wine barrels while at the same time serving as a sound-damping element. One is surrounded by wine. The light is soft as found in a cellar, easy to conserve the wine. Tasting takes place in an intimate space encouraging introspection. Half-light accompanies the visitors along their tour, punctuated by corridors and stairways evoking typically maze-like wine vaults or even the cramped vineyard path. Exploration leads to the self-service enomatic area, to the projection theatre in the basement, and to the conference space on the upper floor. From here, through the mesh of the hanging canvas, the cascade, vineyards and lake are visible – the only place in the structure

from where these may be viewed.

The use of local materials is an integral part of the architectural concept: cement and steel, characteristic elements of the vineyards, along with oak woodwork as present in the winemaking process. Atelier d'architectes Fournier-Maccagnan

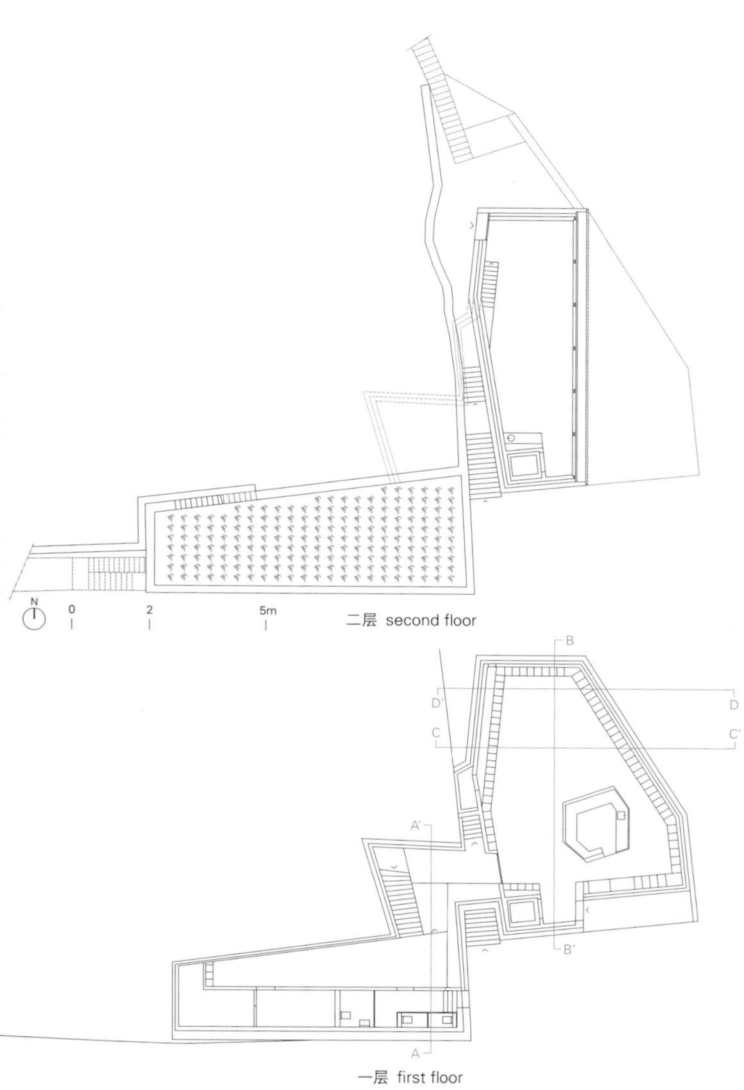

二层 second floor

一层 first floor

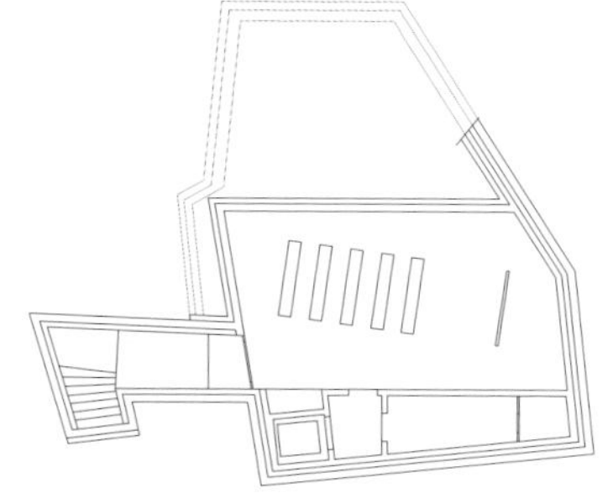

地下一层 first floor below ground

Lavaux Vinorama

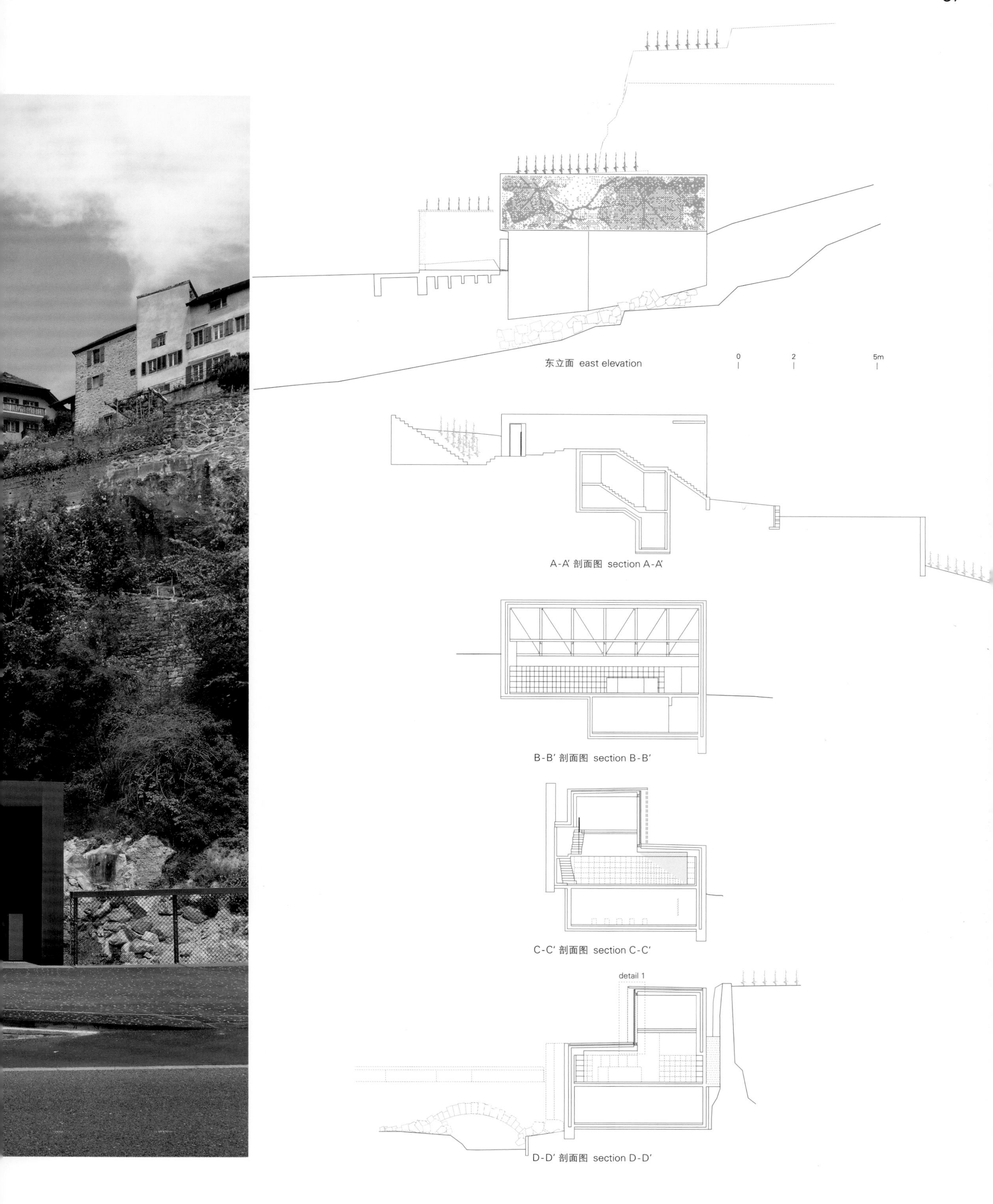

东立面 east elevation

A-A' 剖面图 section A-A'

B-B' 剖面图 section B-B'

C-C' 剖面图 section C-C'

D-D' 剖面图 section D-D'

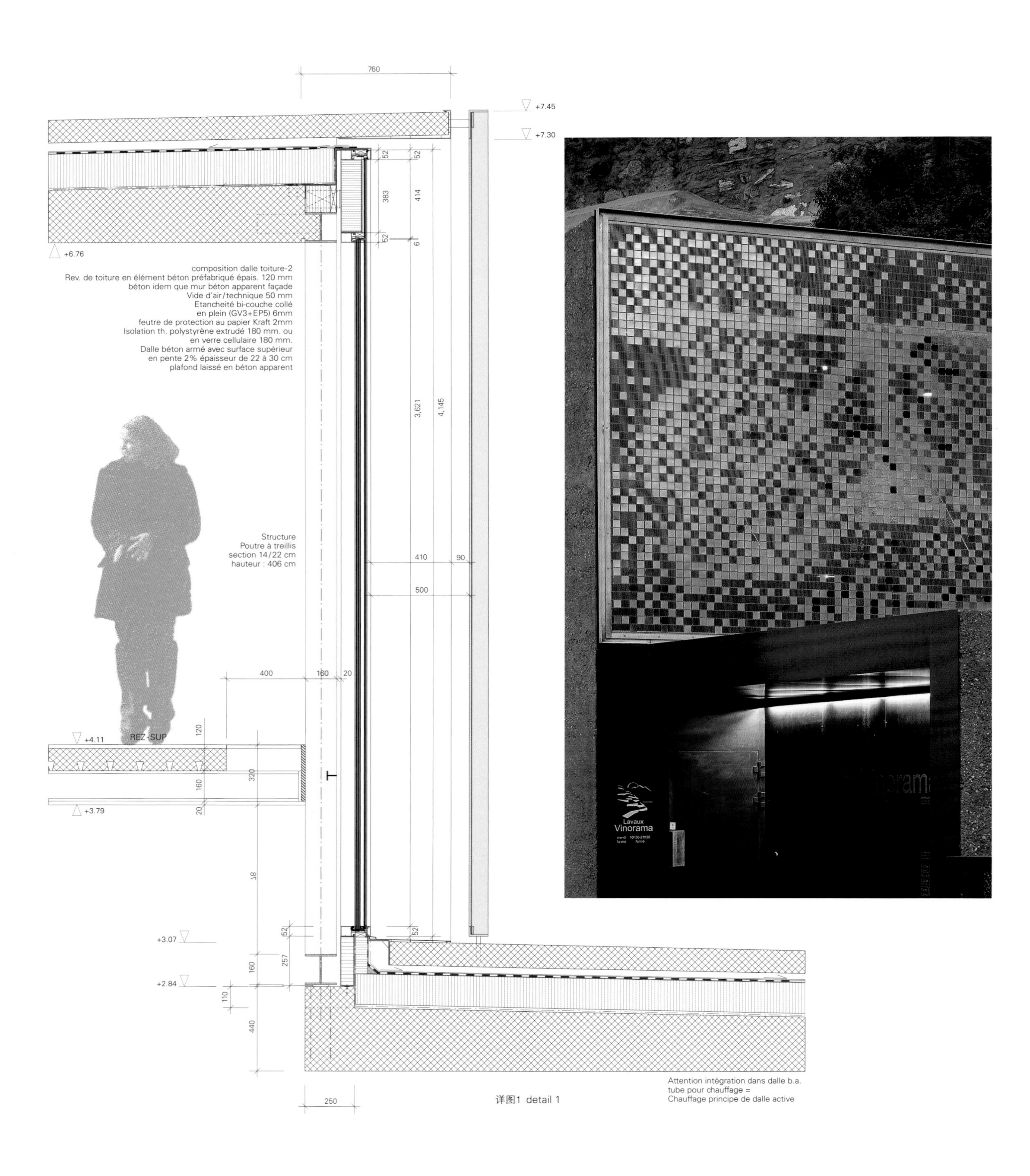
760
+7.45
+7.30
+6.76
composition dalle toiture-2
Rev. de toiture en élément béton préfabriqué épais. 120 mm
béton idem que mur béton apparent façade
Vide d'air/technique 50 mm
Etancheité bi-couche collé
en plein (GV3+EP5) 6mm
feutre de protection au papier Kraft 2mm
Isolation th. polystyrène extrudé 180 mm. ou
en verre cellulaire 180 mm.
Dalle béton armé avec surface supérieur
en pente 2% épaisseur de 22 à 30 cm
plafond laissé en béton apparent
Structure
Poutre à treillis
section 14/22 cm
hauteur : 406 cm
410
90
500
400
160
20
+4.11
REZ-SUP
+3.79
+3.07
+2.84
250
Lavaux
Vinorama
Attention intégration dans dalle b.a.
tube pour chauffage =
Chauffage principe de dalle active

详图1 detail 1

Troll Wall餐厅

Reiulf Ramstad Architects

该建筑是Troll Wall山脚下的一个新标志。毗邻E139的新游客中心建筑是该基地与令人印象深刻的山墙紧密相连的直接产物。这面山墙位于Romsdal山谷，是欧洲最高的竖直悬垂的岩石表面。Romsdal山谷中有一些欧洲最高、最陡峭的悬崖，是一个适合BASE蹦极（包括身着滑翔衣从悬崖跳下的"鸟人"）的深受人们喜爱的地方。该地点为建筑师提供了一块可以建造新服务与信息中心的激动人心的场地。RRA建筑师事务所的提案经过精心规划，与Troll Wall紧密相联。同时，该事物所也是在打造一种特色，使这座建筑成为该区域的一大吸引力。该建筑的规划十分简洁却又不失灵活性，设有一个与众不同的屋顶，该屋顶在周围的壮丽景观中独具特色。这种简单的设计方法赋予该建筑以品质与特色，从而使得服务中心能够在该区域成为人们视线的焦点，以及一座吸引人们来此观光的建筑。

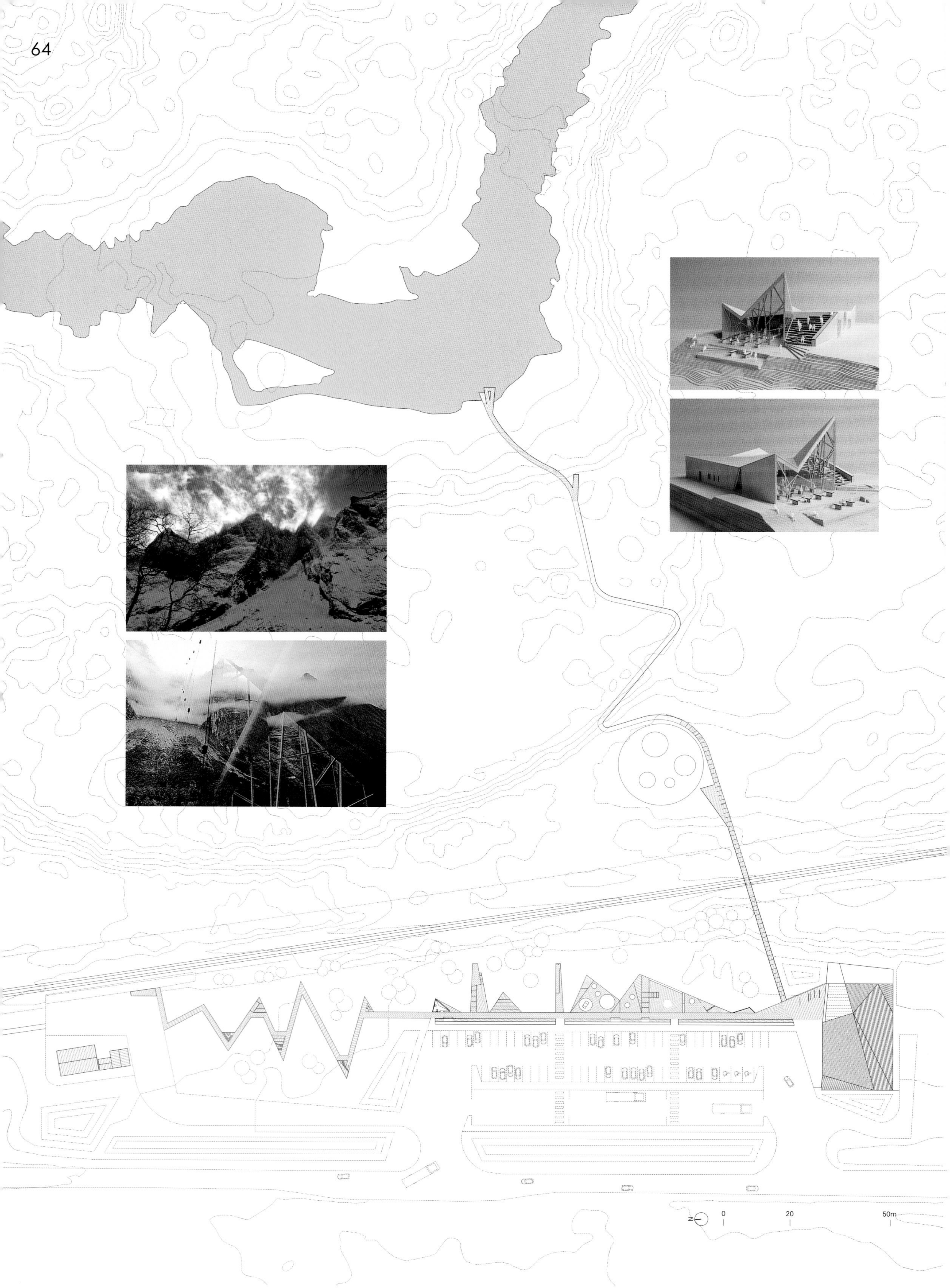
N
0
20
50m

Trollveggen
besøkssenter
Kafeteria / Trollveggen utsikt / Trollveggen film / Souvernir

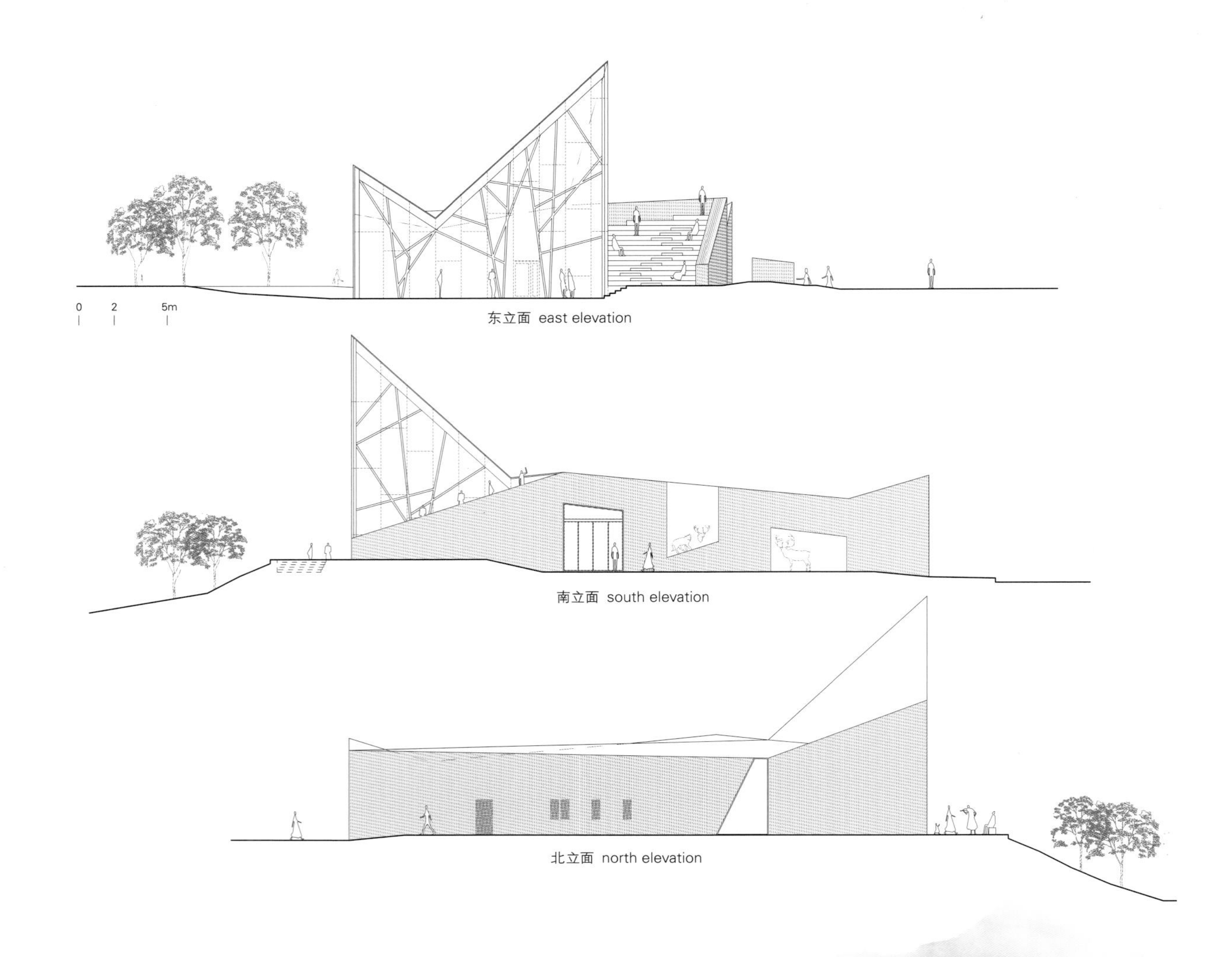
0
2
5m
东立面 east elevation
南立面 south elevation
北立面 north elevation

Troll Wall Restaurant

It's a new cursor at the foot of the Troll Wall. The architecture of the new visitors' centre next to E139 is an outcome of the sites' close connection to the impressive mountain wall, Europe's tallest vertical, overhanging rock face in the Romsdal Valley. The Romsdal valley has some of the tallest, sheerest cliffs in Europe and is a popular place for BASE jumping including "birdmen" jumping off cliffs in Wingsuits. This location allows for an exciting setting for the new service and information centre. RRA's proposal is carefully planned in relation to the Troll Wall. At the same time it is building a character and identity which in itself will be an attraction in the region. The building has a simple, though flexible plan, with a characteristic roof that has its character from the majestetic surrounding landscape. These simple ways of design give the building its character and identity that make the service centre an eye-catcher and an architectural attraction in the region.

Reiulf Ramstad Architects

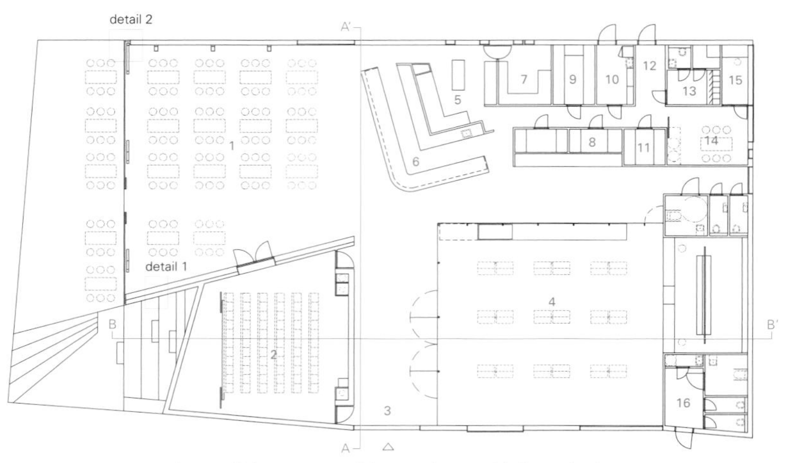

1 餐厅 2 礼堂 3 入口 4 商店 5 厨房 6 服务柜台 7 洗衣房
8 冰柜 9 冷藏室 10 废物存放处 11 储藏室 12 员工入口 13 衣橱
14 员工房间 15 办公室 16 卫生间

1. restaurant 2. auditorium 3. entrance 4. store 5. kitchen 6. serving counter
7. washing 8. fridge 9. freezer 10. waste 11. storage 12. staff entrance
13. wardrobe 14. staff room 15. office 16. toilet

一层 first floor

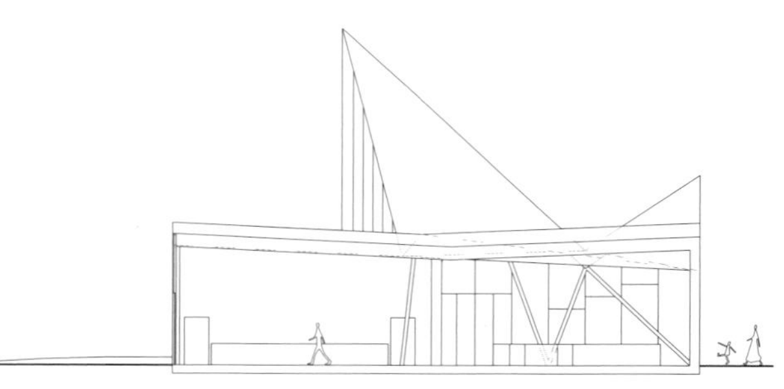

A-A' 剖面图 section A-A'

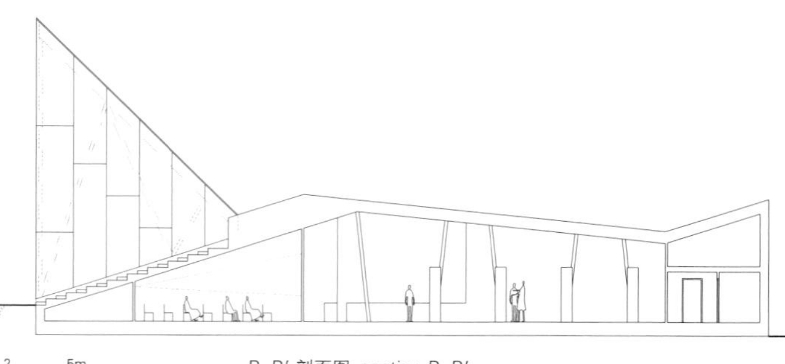

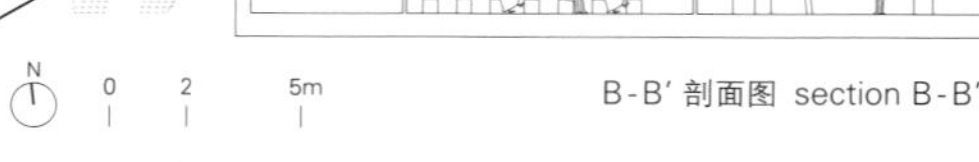

B-B' 剖面图 section B-B'

项目名称：Troll Wall Restaurant
地点：Trollveggen, Møre og Romsdal, Norway
建筑师：Reiulf Ramstad
项目团队：Sunniva N Rosenberg, Christian Fuglset, Ragnhild Snustad, Atle Leira, Espen Surnevik
合作者：Reiulf Ramstad Arkitekter AS
粉刷工：MIR.NO
设备工程师：Tømmerdal Consult AS v / Bjørn Brekke
机械工程师：Nordvest Miljø AS v / ODD Arne Holstad
电气工程师：A. Skare AS v / Roy Dahle
总承包商：M Kristiseter AS
用途：New restaurant and service building
用地面积：2,000m²
建筑面积：700m²
总楼面面积：650m²
设计时间：2009 竣工时间：2011

1. locking pieces(100x50mm) c/c1500mm
2. plywood/veneer plate
3. glazing/sunscreen
4. window profile T-profile in white steel
5. IPE painted on this side(white)
6. connection piece welded to column
7. wood stairs on amfi
8. asphalt membrane
9. 24x48 batten
10. sealing tape
11. rafters 48x198
12. column 100x100
13. 48x48 batten
14. stud wall 18mm OSB
15. acoustic cloth
16. white preforated plywood
17. air chamber
18. paneling

详图1 detail 1

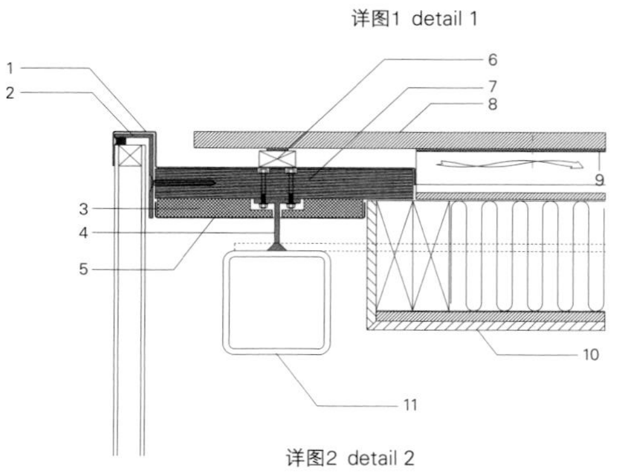

1. seizure in black zinc glued to glass
2. steel angle 100x50 white varnish
3. 23mm adhesive xps
4. T-profile 50x50 white varnish welded to white HUP
5. white fitting
6. plate cladding
7. 2x20mm (40mm) veneer
8. paneling
9. sealing tape
10. white veneer cladding
11. white HUP

详图2 detail 2

轻井泽别墅A

Satoshi Okada Architects

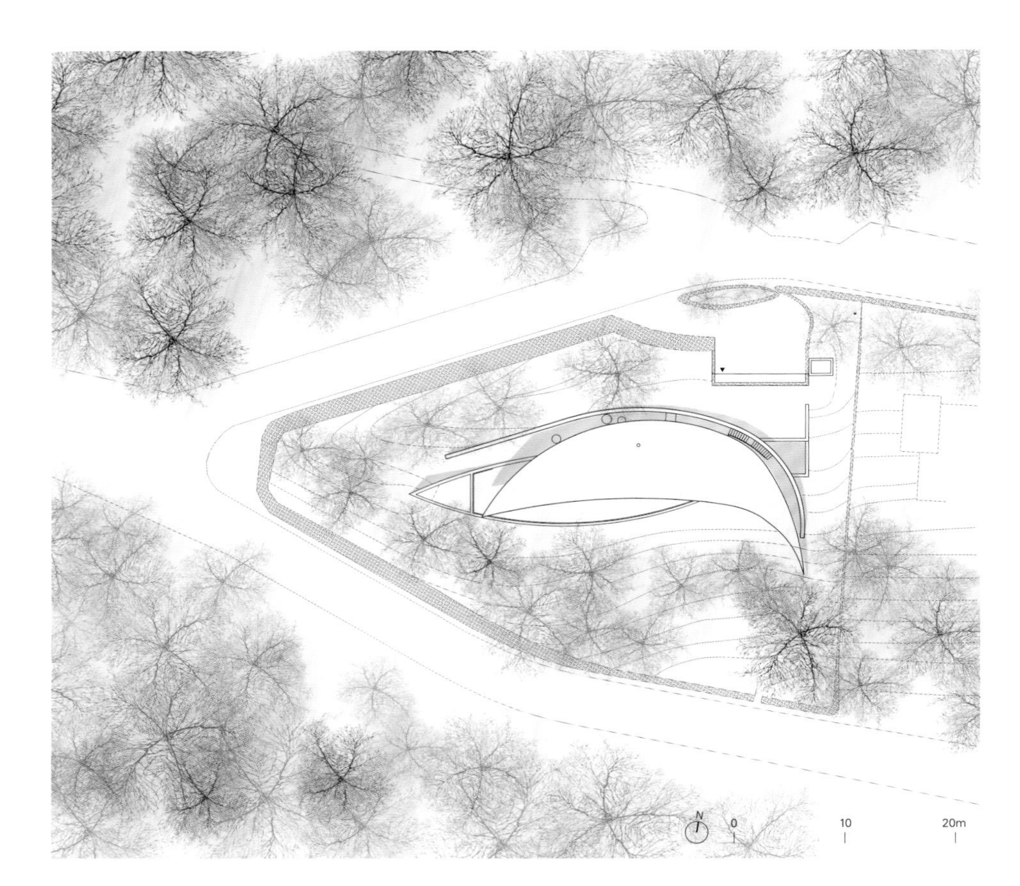

轻井泽别墅A是为长野县东边的轻井泽镇中的一座度假住宅设计的。它高出海平面1000m，自古以来都是最受人们欢迎的夏季度假胜地之一，该地与东京之间的交通也十分便捷。类似于三角形的基地朝南向下倾斜，在这里可以越过一座山谷远眺到山脊的自然美景。山上种有大量古老且高大的落叶性树木，而山下则满是竹草。但是该场地东边的附近坐落着一座邻里住宅，它也是该地最后一个碍眼的人工元素。根据该地区的规章制度，任何类型的屋顶都应该有1/5的倾斜度，同时建筑应至少距离公共道路10m，并且不得多于两层。

该别墅由两部分组成。一部分是半地下的私人房间——客房、卧室、图书室以及一个带有洗衣房的浴室，这部分是由钢筋混凝土结构打造而成的，向南面的区域开放。该项目细长的平面形状极其适合已有的地理轮廓线，同样也适合于用来抵抗地面水平载荷的承重墙。另一部分是位于地上的公共领域建筑——起居室、餐厅以及厨房。这部分由钢结构构成，完全由透明的玻璃墙体围合起来。因为车库和入口都位于地下的山脊线上，因此后者看起来是独立存在的。

弯曲的屋顶是这座别墅最特别的元素，它意在保护上层建筑，就好像一只将翅膀完全伸展开为了在巢中进行孵化的鸟儿，它还能够柔和地遮挡住来自东边邻近建筑的视线。在计划中，它是按照如下步骤设计的：首先，将金属板弯曲成二维效果；然后，将其倾斜成1/5的倾斜度；最后，将金属板切割开，以覆盖住建筑。在结构方面，屋顶是由科尔顿钢板的夹芯板构成的，其中设有格状肋，屋顶总体厚度仅为15cm。

在这个项目中，建筑师们与英国园艺家Paul Smither（如今活跃在日本）共同合作，尝试将建筑与景观融合在一起，将项目演变成一处整体环境。

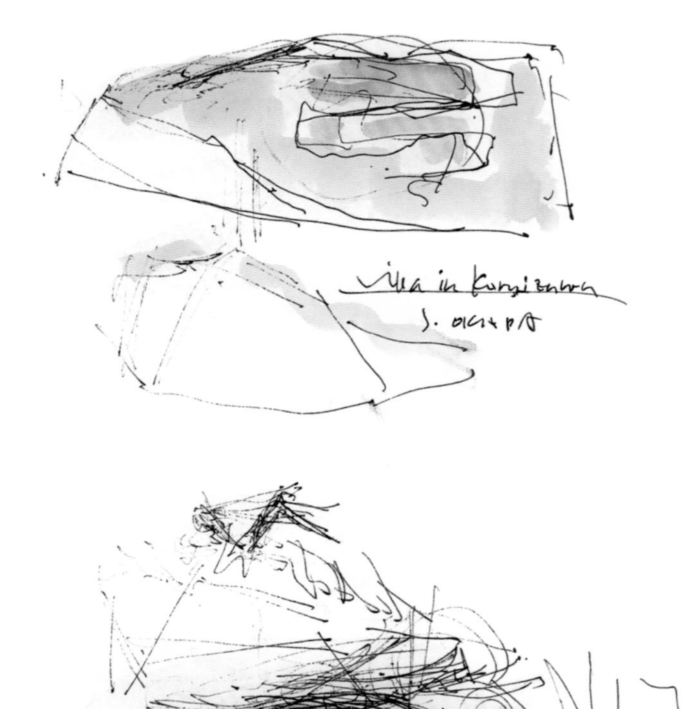

Villa A in Karuizawa

Villa A in Karuizawa is designed for a vacation house in the town of Karuizawa at the east end of Nagano Prefecture. It is about 1000 meters above sea level, historically one of the most popular summer resorts within easy reach from Tokyo. The quasi triangular site is sloping down towards the south, from which a beautiful nature view can be appreciated over a valley to mountain ridges. There are plenty of old and tall deciduous trees above and overall bamboo grasses below. However, a neighbor house is close to the east border, which is the last unfavorable artificial element seen from the site. According to regulations in the district, at least 1/5 gradient is required to any kind of roof, building must be set at least 10 meters off the public road borders, and the building should be less than two-story.

The villa is composed of two portions. One is the semi-underground part for private rooms – guest room, bedrooms, library, and a bathroom with laundry – made of reinforced concrete

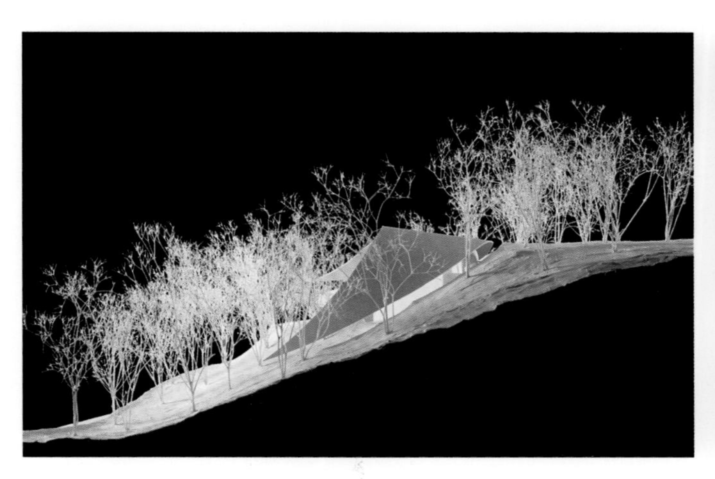

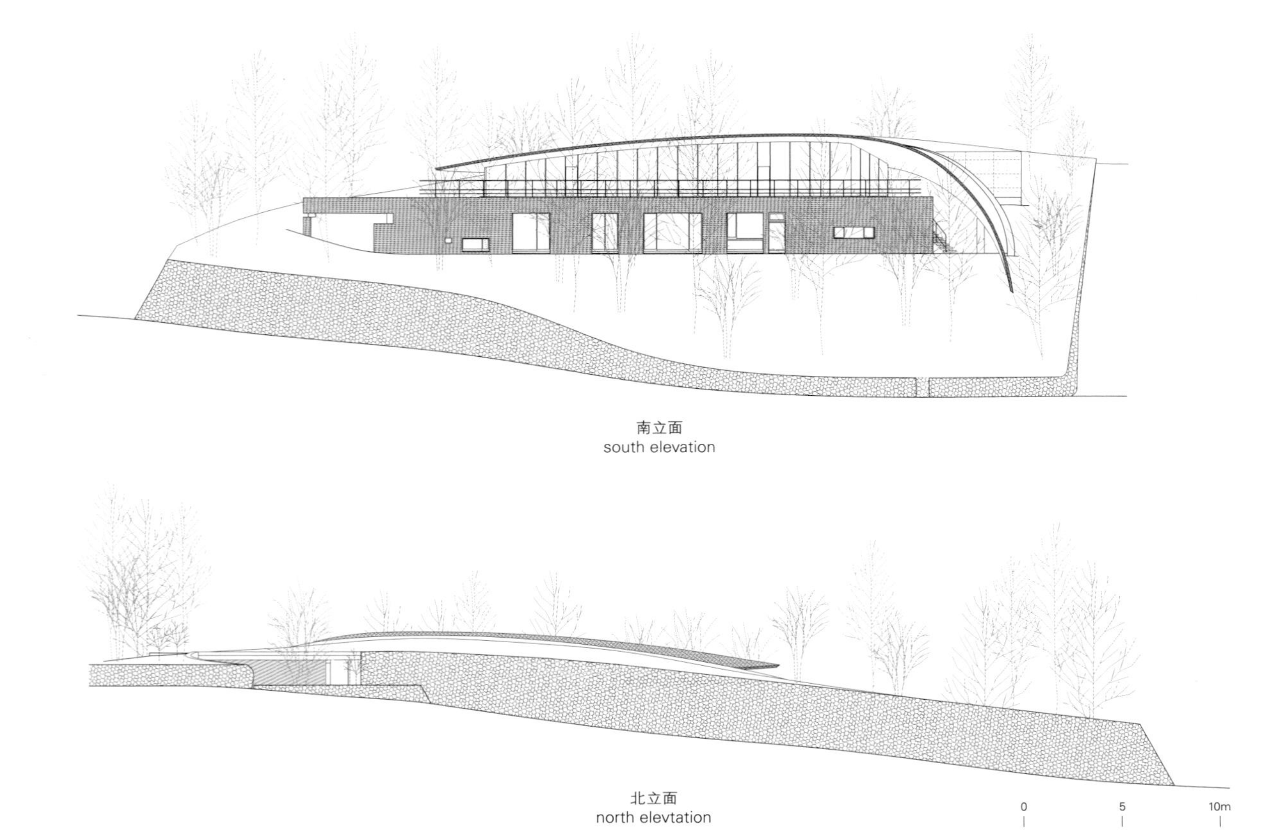

南立面
south elevation

北立面
north elevtation

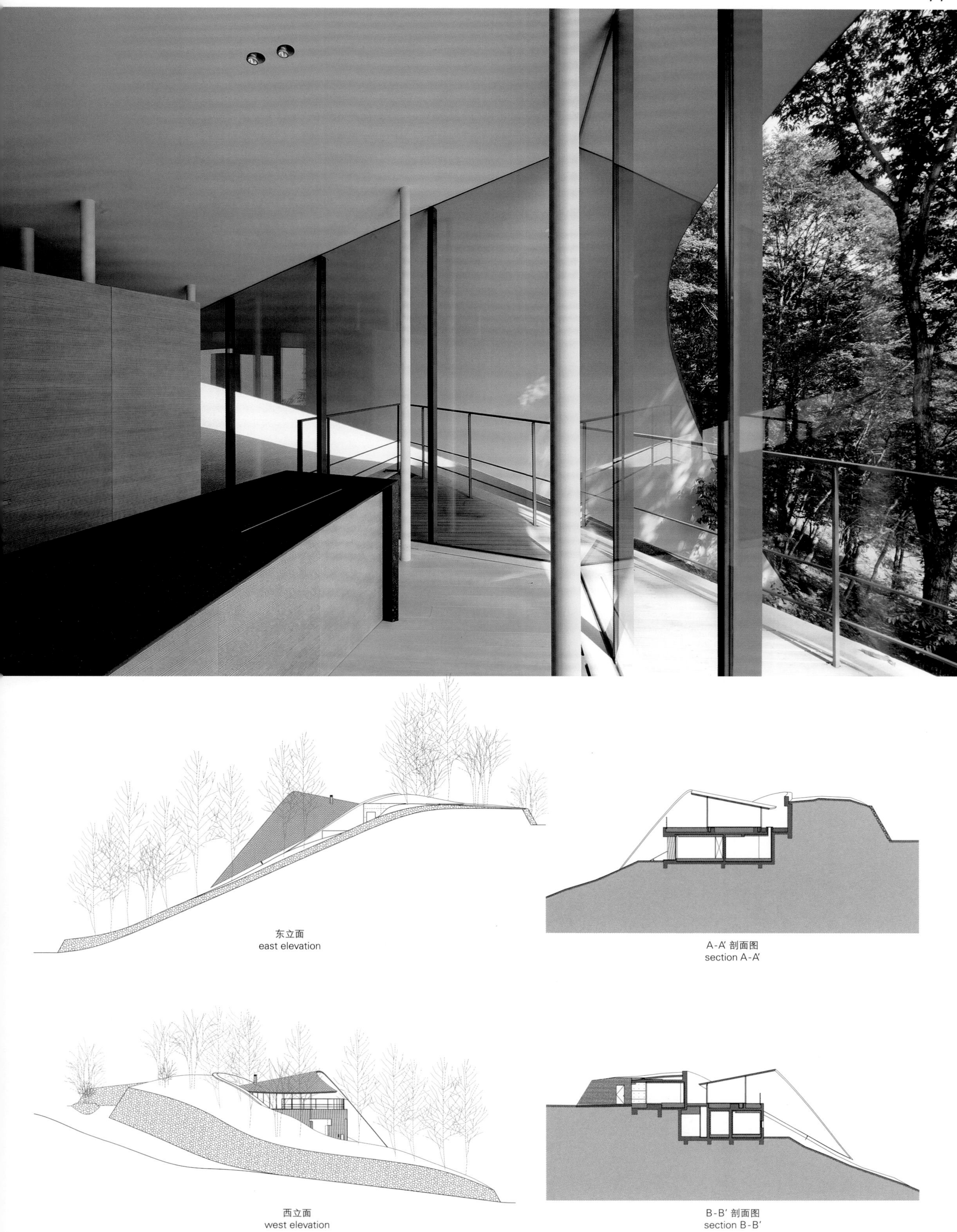

东立面
east elevation

A-A' 剖面图
section A-A'

西立面
west elevation

B-B' 剖面图
section B-B'

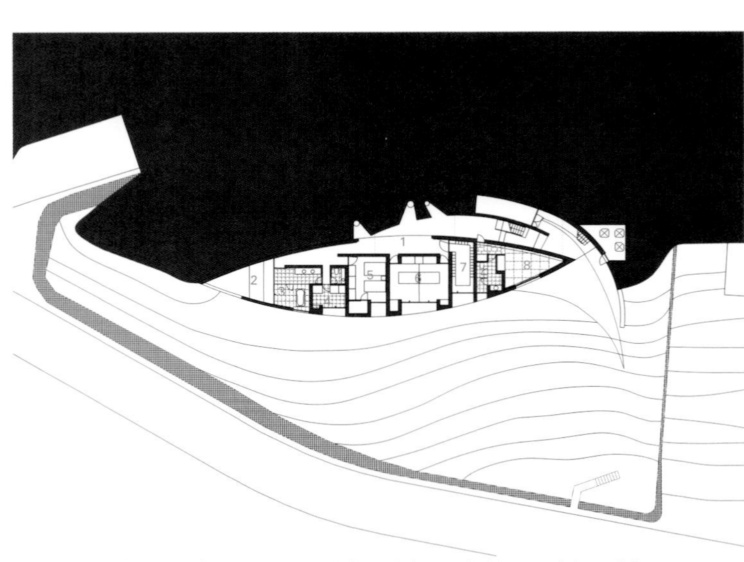

1 大厅 2 露台 3 浴室 4 公共设施 5 主卧 6 儿童卧室 7 图书室 8 客房
1. hall 2. terrace 3. bathroom 4. utility 5. master bedroom
6. children's bedroom 7. library 8. guest room

地下一层 first floor below ground

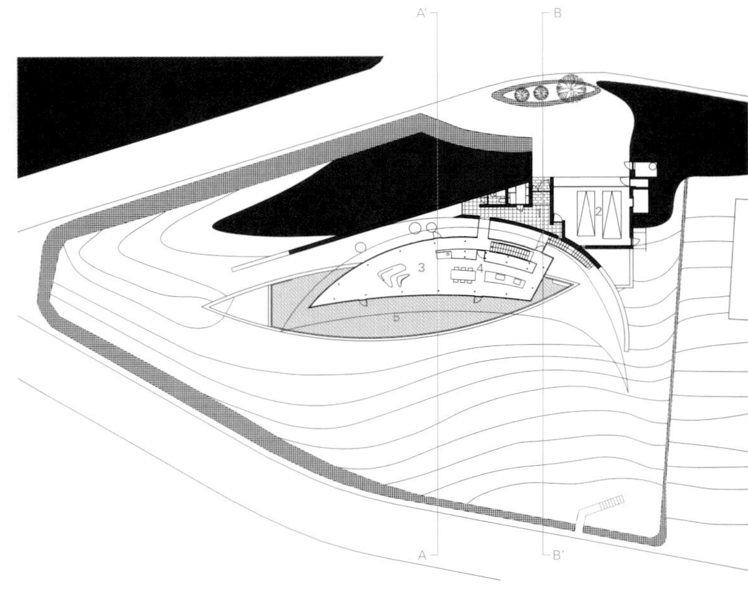

1 入口 2 车库 3 起居室 4 餐厅&厨房 5 露台
1. entrance 2. garage 3. living room 4. dinning & kitchen 5. terrace

一层 first floor

项目名称：Villa in Karuizawa (Villa A)
地点：Karuizawa, Nagano Prefecture, Japan
建筑师：Satoshi Okada
项目团队：Isao Kato, Lisa Tomiyama, Satoshi Okada architects
结构设计师：Hirokazu Toki
甲方：Hiroyuki Ando
用地面积：2316.58m^2
建筑面积：424.44 m^2 总楼面面积：434.63m^2
结构：steel structure on reinforced concrete structure
材料：marble, granite, lava stone, wood flooring, tatami-mats, etc.
设计时间：2009.4 — 2009.8
竣工时间：2011.6
摄影师：©Hiroshi Ueda(courtesy of the architect)-p.74~75, p.76, p.77, p.79, p.80 bottom
©Sergio Pirrone-p.72, p.78, p.80 top

structure, open to the south side of terrain. The spindle plan shape is the most suitable for existing contour lines geographically as well as for the bearing wall against the horizontal load from the earth. The other is an overground building for the public realm – living, dining, and kitchen. It is consisted of steel structure totally surrounded by transparent glass walls. Because garage and entrance are placed underground on a ridge line, the latter one looks independent.

A bent roof, the most particular element in this villa, was intended to shelter the upper building with an image of a bird's wing fully extended for brooding over a nest. It also functions to blind softly the neighbor building on the east. Schematically, it was designed as follows: first to bend a plate two dimensionally, second to incline at 1/5 gradient, and at last to cut-out the plate for covering the building. Structurally, the roof is made of sandwiched panels of Corten steel plate with lattice ribs inside, and it is only 15cm thick in total.

In this project, architects try to integrate architecture and landscape into designing a whole environment in cooperation with Paul Smither, British horticulturist, active in Japan today.

Satoshi Okada Architects

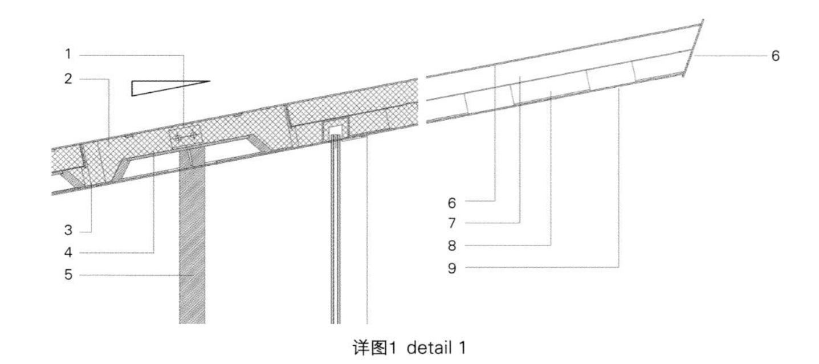
详图1 detail 1

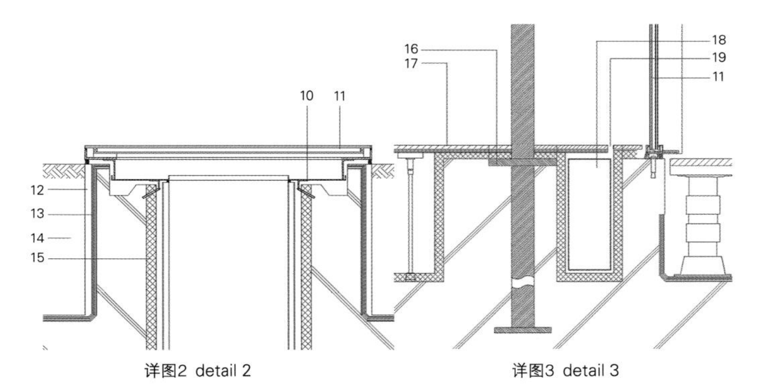
详图2 detail 2

详图3 detail 3

1. steel piece for erection: L-65x65x6
2. steel rib plate: 6mm
3. steel plate: 4.5mm acrylic resin paint finish
4. steel plate: 6mm ø480mm
5. column: round steel ø80mm
6. Corten steel plate: 4.5mm
7. steel deck
8. steel rib plate: 4.5mm L=250 @400
9. steel plate: 4.5mm fluoric resin paint finish
10. condensation gutter: steel plate 2.3mm
11. double glazing: 6+12+8mm
12. mortar 20mm
13. double layer sealing membrane: 15mm
14. concrete: 150mm
15. thermal insulation: 30mm
16. steel base plate: 200x200x25mm
17. wood flooring/floor heating system
18. A/C trench
19. A/C supply slit

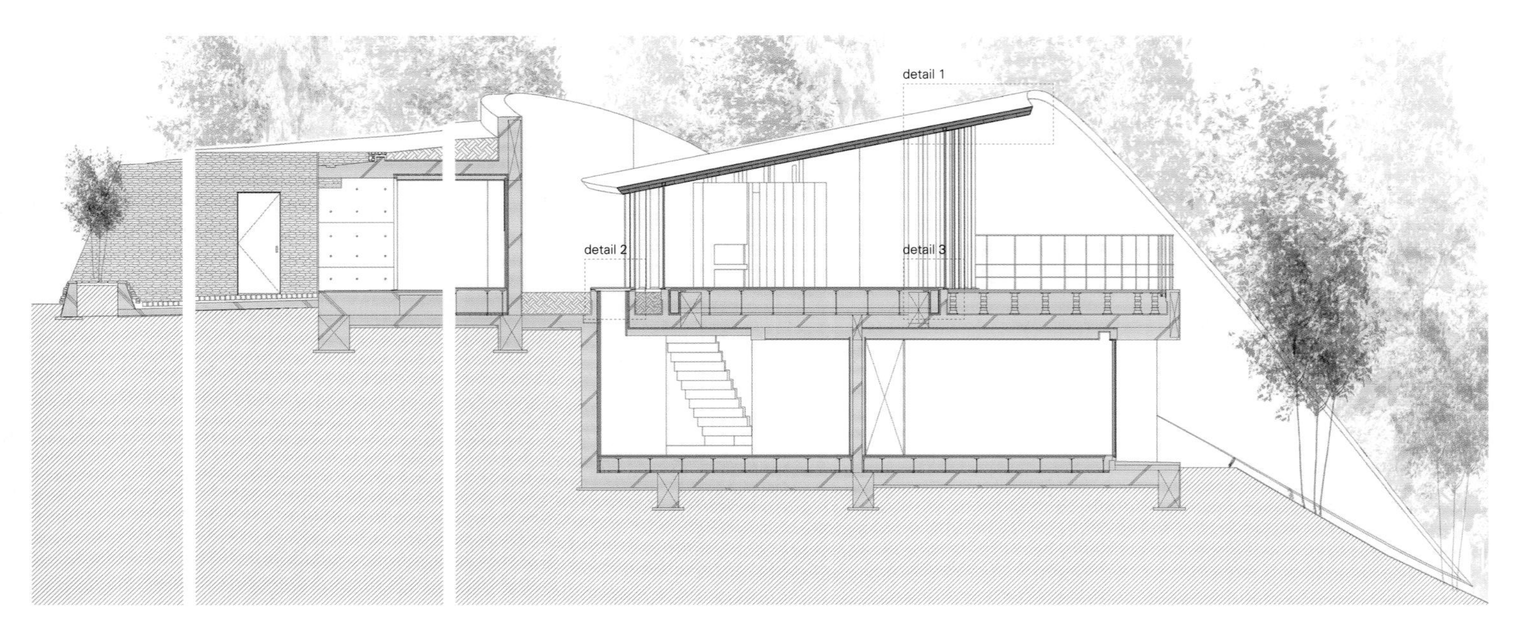

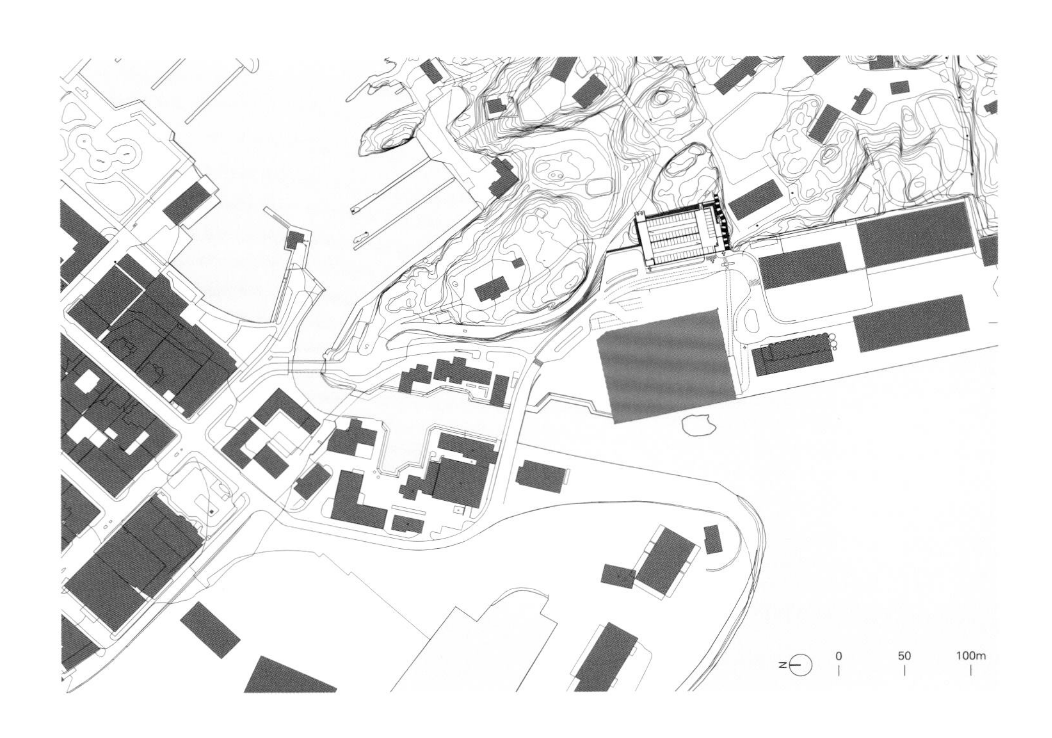

Kilden表演艺术中心

ALA Architects

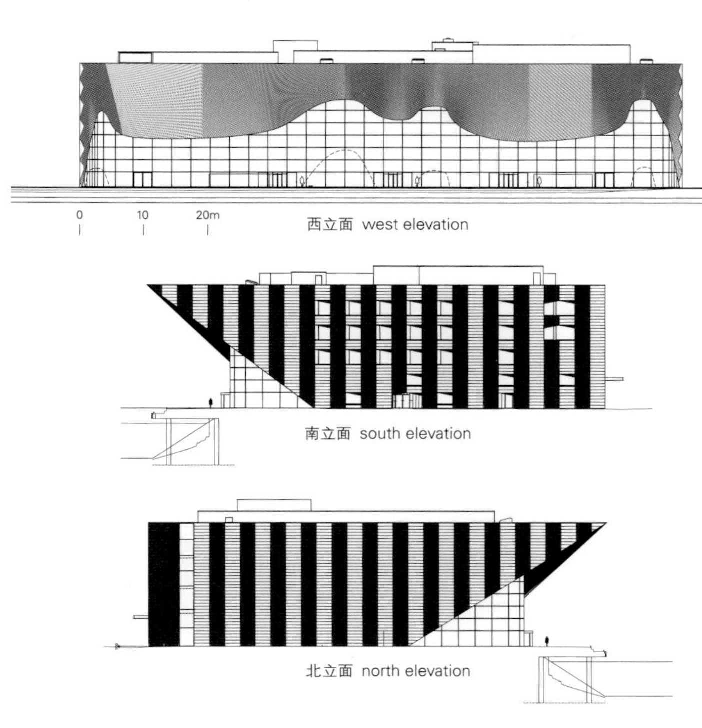

西立面 west elevation

南立面 south elevation

北立面 north elevation

Kilden表演艺术中心容纳了克里斯蒂安桑城里所有的表演艺术机构。鞋盒形状的音乐厅专为交响乐团建造，可以容纳1200名观众。Agder郡剧院有一个设有700个座位的剧院大厅，它拥有适于戏剧演出的极佳音响效果和宽敞的舞台结构，同时还设有一个剧院塔楼以及位于舞台下方的机械设备。该剧院大厅还可以经过调整，用来举办音乐戏剧表演，并用来展示当地歌剧公司所创造的作品。除了主厅，kilden表演艺术中心还有一个为临时剧院而设的演播室舞台和为各类活动准备的多功能大厅。

Kilden表演艺术中心给观众带来了不同的体验。该项目的基本建筑内容就是将音乐厅、剧院和多功能大厅与支撑空间相结合。各大厅根据它们各自的功能被设置成一排，服务区域的布局方式使得公共生产设备能够不受任何干扰且有效地服务于所有大厅。该服务区域位于大厅的一侧，公共门厅则位于另一侧。

礼堂下方塑造出的悬臂式外观覆有一面连续的弯曲木质墙体。公共门厅和入口区域的立柱也覆有这种平滑的结构。由当地橡木打造而成的弯曲墙体将人们的日常生活与大厅内部艺术表演的虚幻世界划分开来。门厅上方的这种超凡形状吸引观众和公众聚集在这座建筑之下，并寓意着它背后所展示的艺术与表演的丰富性和多样性，也代表着艺术机构互相合作的潜力。弯曲墙体由实心橡木板制成，并运用了由电脑辅助的建模技术。它是一个真实的、可触摸的实体形式，而不仅仅是一层薄薄的材质。这种木质结构也改善了门厅中的音响效果。

该建筑的其他三面均覆有单一的黑色铝质立面，完善并强调了由门厅墙体创造出的华丽空间。整座建筑仿佛是为其所容纳的“乐器”——表演大厅——而量身定做的高级盒子。音乐厅是一个黑色的鞋盒形状的体量，越接近天花板，越消失在人们的视野之中，并创造出回荡的音响效果。音乐厅的墙壁上覆盖着连续的三角面黑色纤维混凝土构件，创造出最佳的音效调节表面。观众和表演者被轻质优雅的木质表层所环绕，打造出了一种类似于家具的明亮构件漂浮于巨大、黑暗的大厅之中的效果。塑造木质构件的形状可以创造出环绕在观众和表演者周围的、更精准的音效调节表面。

剧院大厅拥有一种较为随意和实验性的特质。这种氛围促使观众沉浸在表演当中。大厅的属性和感觉可以进行调整，以迎合不同的活动和表演。较小的音乐厅有它们自己的特征以及独特的设计手法。

Kilden表演艺术中心的建筑将克里斯蒂安桑城表演艺术机构的努力结合在一起并加以强化，还吸引了更多的观众来观看演出。这些功能相结合所产生的效果远远高出各个部分相加之和，即建筑本身的效果。

Kilden Performing Arts Centre

Kilden gathers all the performing arts institutions of Kristiansand under one roof. A shoebox concert hall with a capacity of 1200 seats is built for the symphony orchestra. Agder county theatre will perform in a 700-seat theatre hall with excellent drama acoustics and extensive stage mechanics with a theatre tower and understage machinery. The theatre hall is adjustable for music theatre performances, and will also house productions by the local opera company. In addition to the main halls, Kilden also includes a studio stage for contemporary theatre and a multipurpose hall for a wide range of activities.

Kilden produces experiences. The fundamental architectural content is to make the concert, theatre and multipurpose halls work together with the support spaces. The halls arranged in a row according to their functions, and the organization of the service areas allows common production equipment to serve all halls efficiently and without interference. The service areas are located on one side of the halls and the public foyer on the other.

The cantilevering shapes created by the undersides of the auditoriums are clad with a continuous curving wooden wall. This

HESTMANDEN

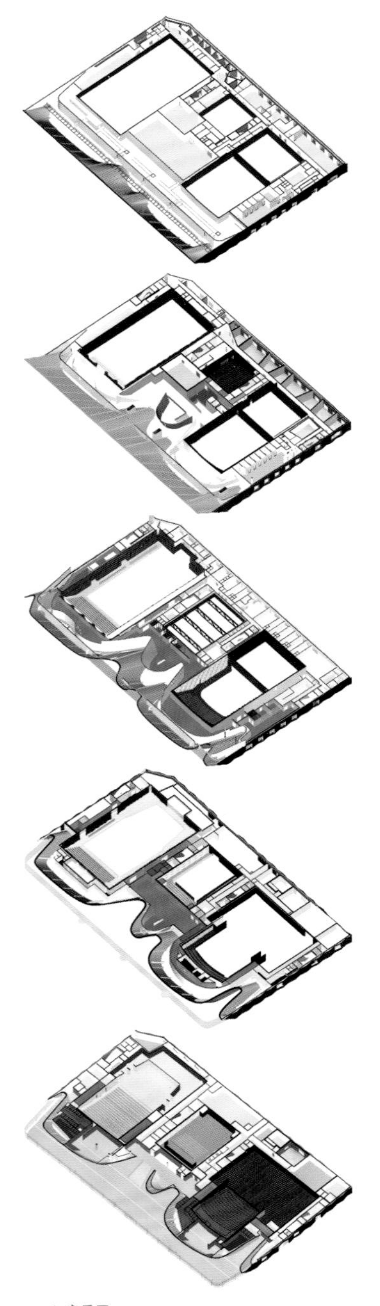

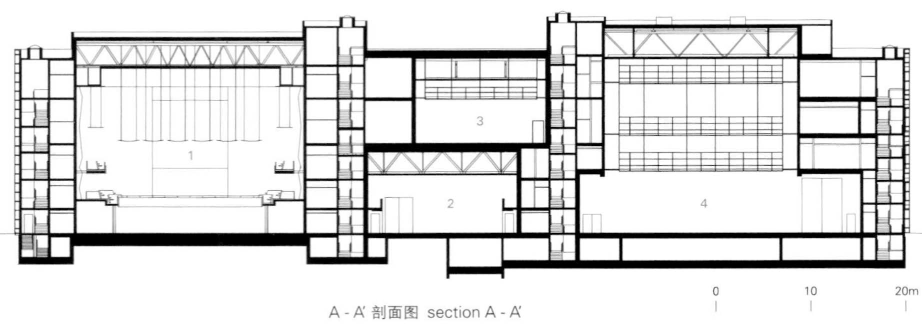

1 音乐厅
2 多功能大厅
3 黑色盒状区域
4 剧院大厅
5 门厅
6 支撑空间

1. concert hall
2. multipurpose hall
3. black box
4. theatre hall
5. foyer
6. support spaces

0 10 20m

A - A' 剖面图 section A - A'

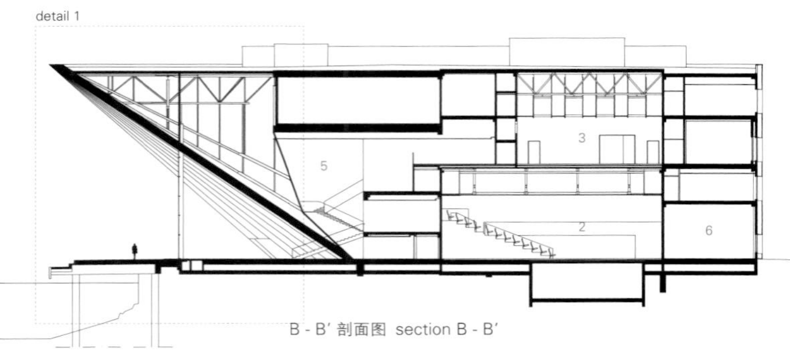

B - B' 剖面图 section B - B'

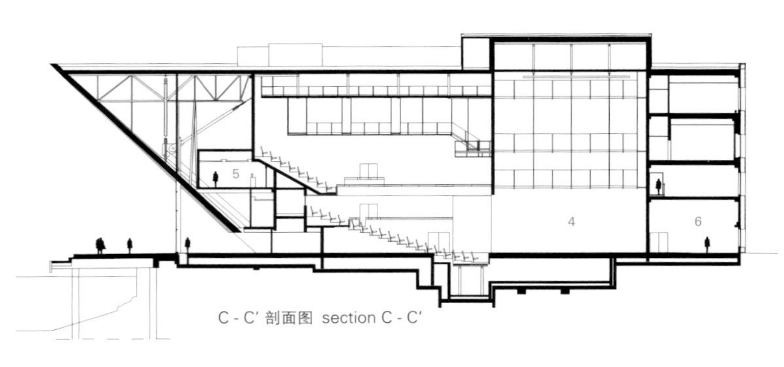

C - C' 剖面图 section C - C'

项目名称：Kilden Performing Arts Centre
地点：Kristiansand, Norge
建筑师：ALA Architects ltd
项目团队：Juho Grönholm, Antti Nousjoki, Janne Teräsvirta, Samuli Woolston, Niklas Mahlberg
合作商：SMS Arkitekter AS
项目管理：Faveo
结构工程师：Multiconsult AS
机械工程师：Sweco Groner
电气工程师：COWI
照明设计：Julle Oksanen
主要合作商：AF Gruppen Norge AS
甲方：Teater- og Konserthus for Sørlandet IKS
用途：Theatre and concert hall
总楼面面积：16,000m²
总体积：128,000m³
设计时间：2004—2005
竣工时间：2011
摄影师：©Kjartan Belland(courtesy of the architect)-p.92
©Tuomas Uusheimo(courtesy of the architect)-p.86, p.88, p.93, p.94, p.95
©Hufton+Crow-p.82, p.84~85, p.87, p.89 (except as noted)

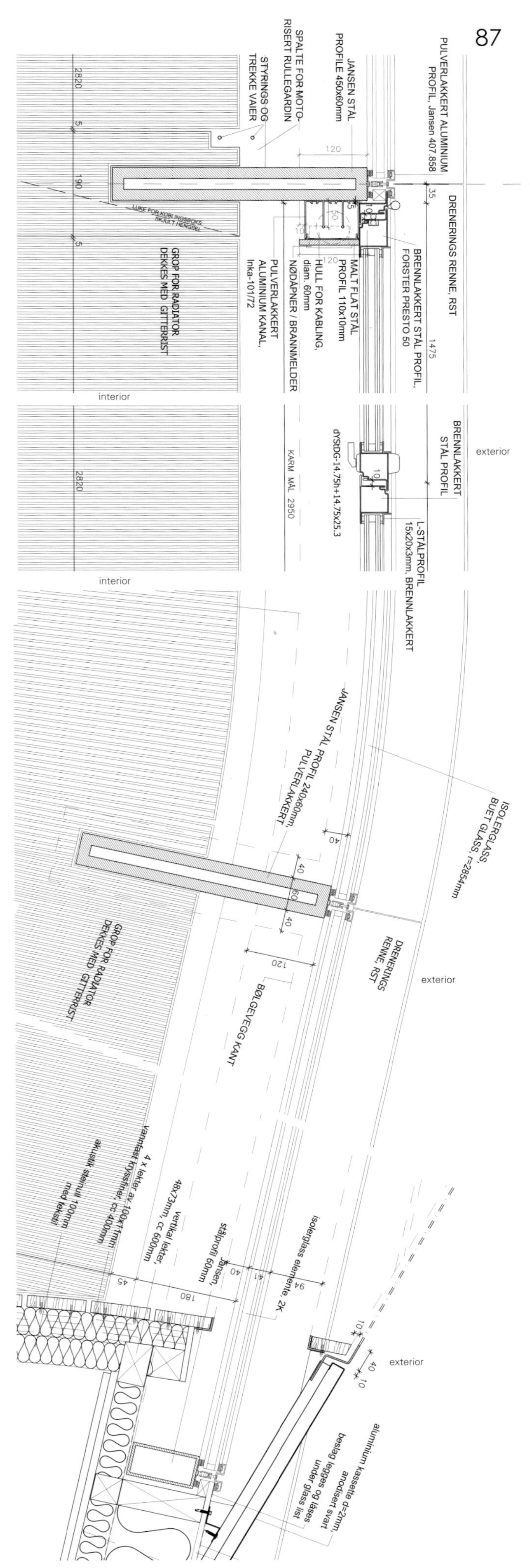

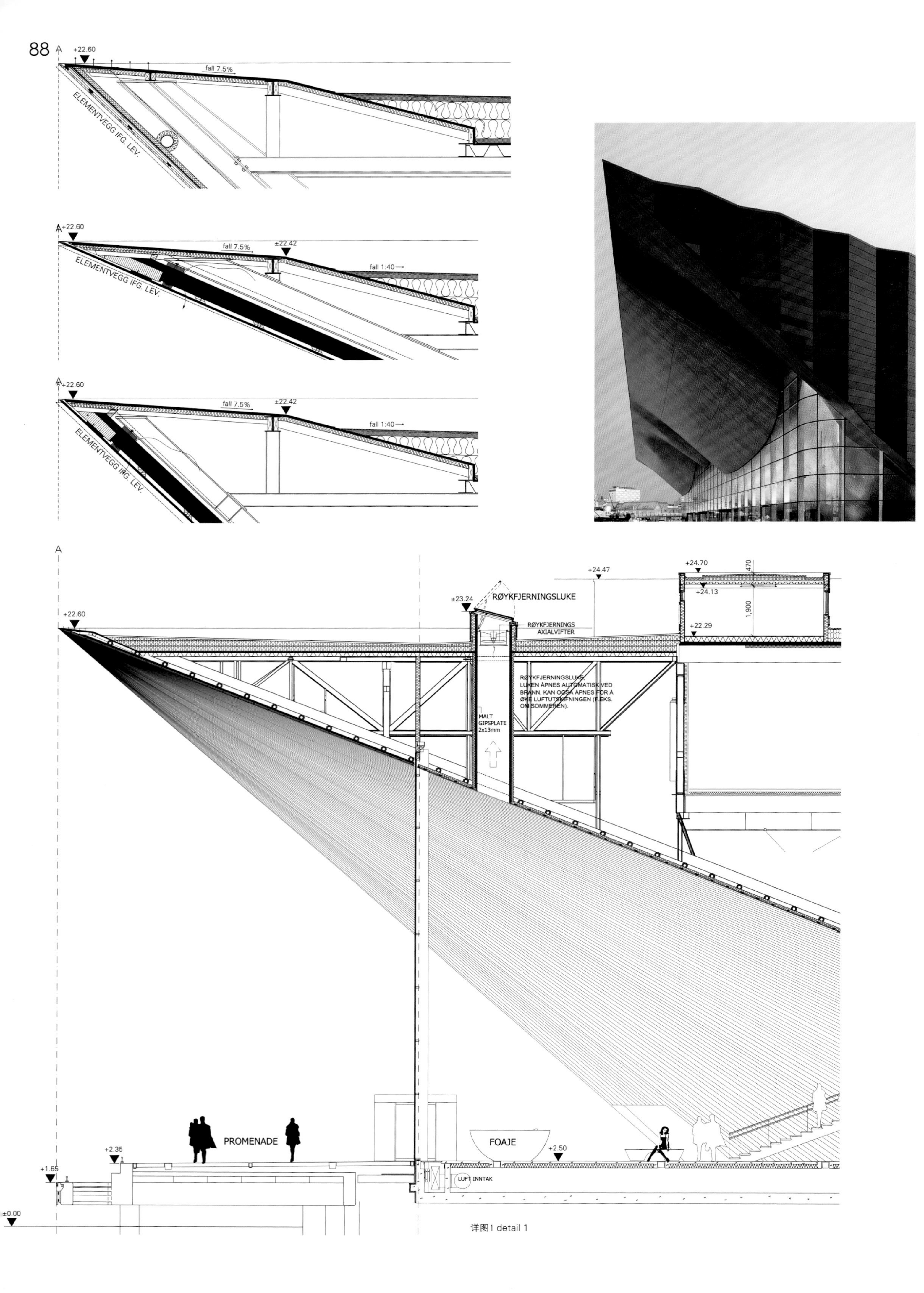
A
+22.60
fall 7.5%
ELEMENTVEGG IFG. LEV.
A
+22.60
fall 7.5%
±22.42
fall 1:40
ELEMENTVEGG IFG. LEV.
A
+22.60
fall 7.5%
±22.42
fall 1:40
ELEMENTVEGG IFG. LEV.
A
+22.60
±23.24
RØYKFJERNINGSLUKE
+24.47
+24.70
470
+24.13
1.900
+22.29
RØYKFJERNINGS
AXIALVIFTER
RØYKFJERNINGSLUKE,
LUKEN ÅPNES AUTOMATISK VED
BRANN, KAN OGSÅ ÅPNES FOR Å
ØKE LUFTUTSKIFNINGEN (F.EKS.
OM SOMMEREN).
MALT
GIPSPLATE
2x13mm
PROMENADE
FOAJE
+2.35
+2.50
+1.65
LUFT INNTAK
±0.00

详图1 detail 1

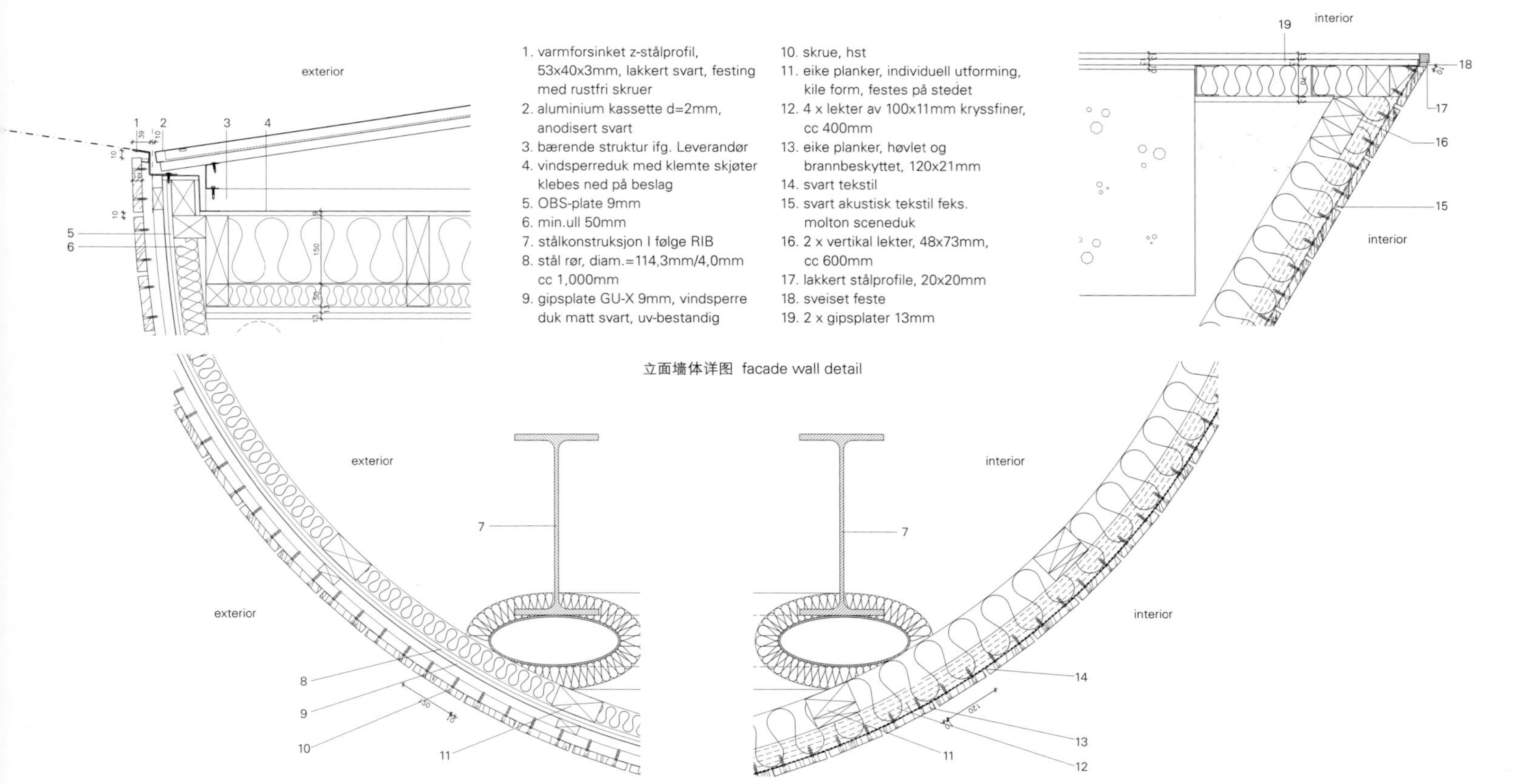

立面墙体详图 facade wall detail

©Hufton+Crow(courtesy of the architect)

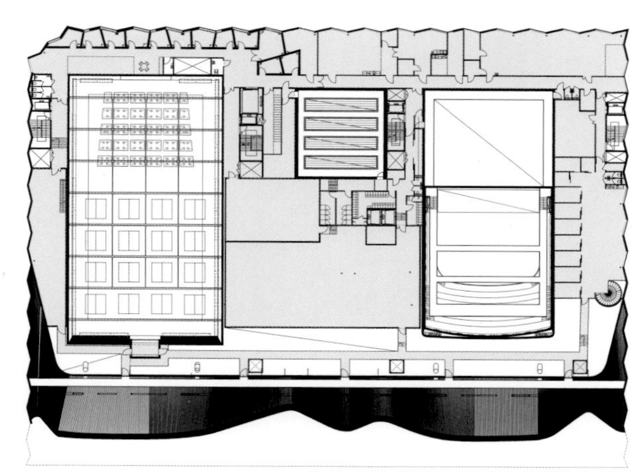

四层 fourth floor

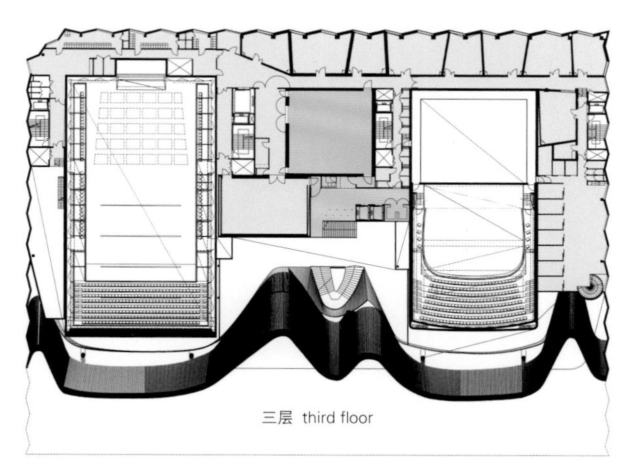

三层 third floor

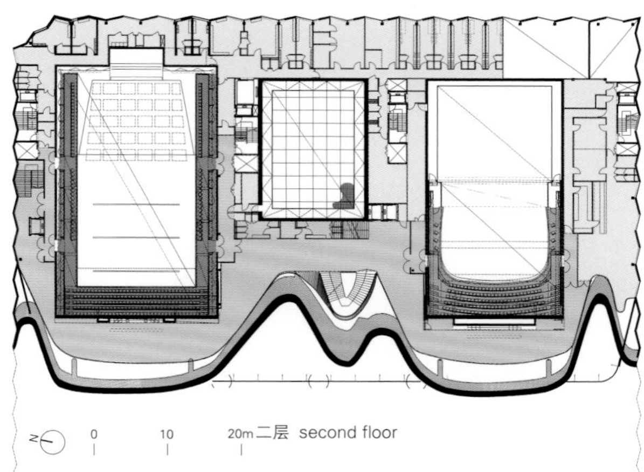

二层 second floor

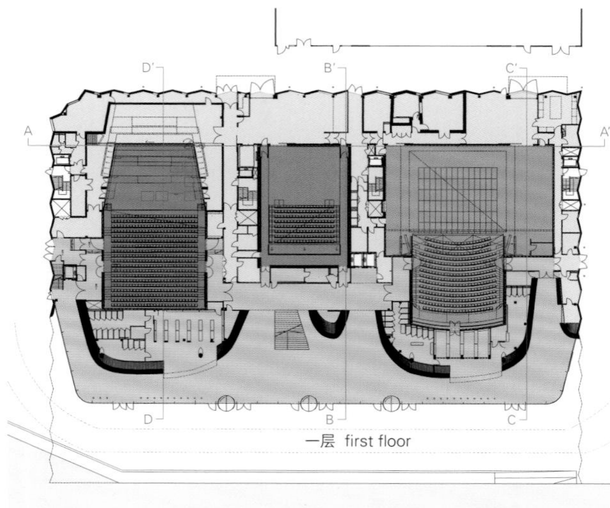

一层 first floor

前厅 foyer
音乐厅 concert hall
多功能大厅 multipurpose hall
剧院大厅 theatre hall
黑色盒状区域 black box
支撑空间 support spaces
餐厅 restaurant

smooth form covers the public foyer and the entrance area on the pier. The curving wall, which is made of local oak, divides the everyday life from the fantasy world of the arts and performances inside the halls. The dramatic shape rising over the foyer attracts the audiences and the public to gather under the building, signifying the richness and diversity of arts and performances behind it, and representing the potential of the arts organizations working together. The curving wall is manufactured from solid oak planks with the help of computer aided modeling techniques. It is a real, tactile, physical form, not just a thin layer of texture. The wooden shape also improves the acoustic conditions of the foyer.

The monotonous, black aluminum facades cover the three other sides of the building, complementing and emphasizing the spectacular space created by the foyer wall. The building is like a high-class case for the "instruments" – the performance halls – inside. The concert hall is a dark, shoebox-type volume which visually disappears towards the ceiling, creating the reverberating volume for the acoustics. The walls of the concert hall are clad with continuous, triangularly faceted black fiber concrete elements, which create acoustically optimal surface modulation. The audience and the performers are surrounded by light, elegant wooden surfaces, which create a bright furniture-like element floating in the vast, dark hall. The more delicate acoustical surface modulation around the audience and the performers is created by shaping this wooden element.

The theatre hall has a more casual and experimental character. It encourages the viewer to become immersed in the performances. The nature and feel of the hall can be adjusted to suit different events and performances. The smaller halls have their own characters and individual design approaches.

The architecture of Kilden combines and consolidates the efforts of the Kristiansand performing arts institutions and attracts the audience to witness them. Together these functions amount to more than the sum of their parts – that is architecture. ALA Architects

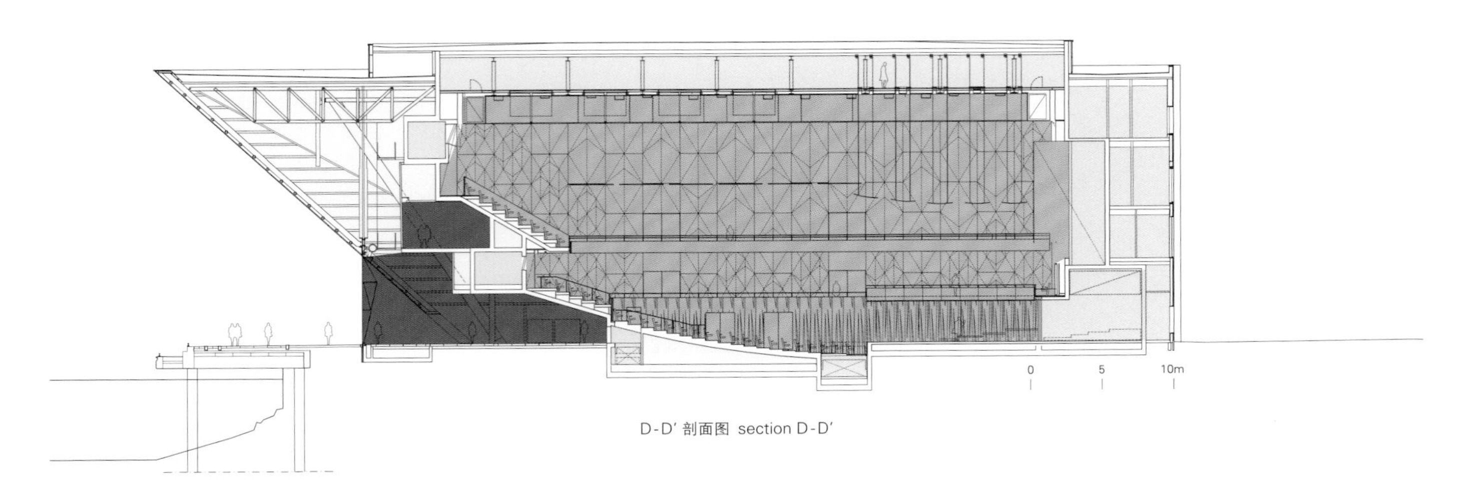

D-D′ 剖面图 section D-D′

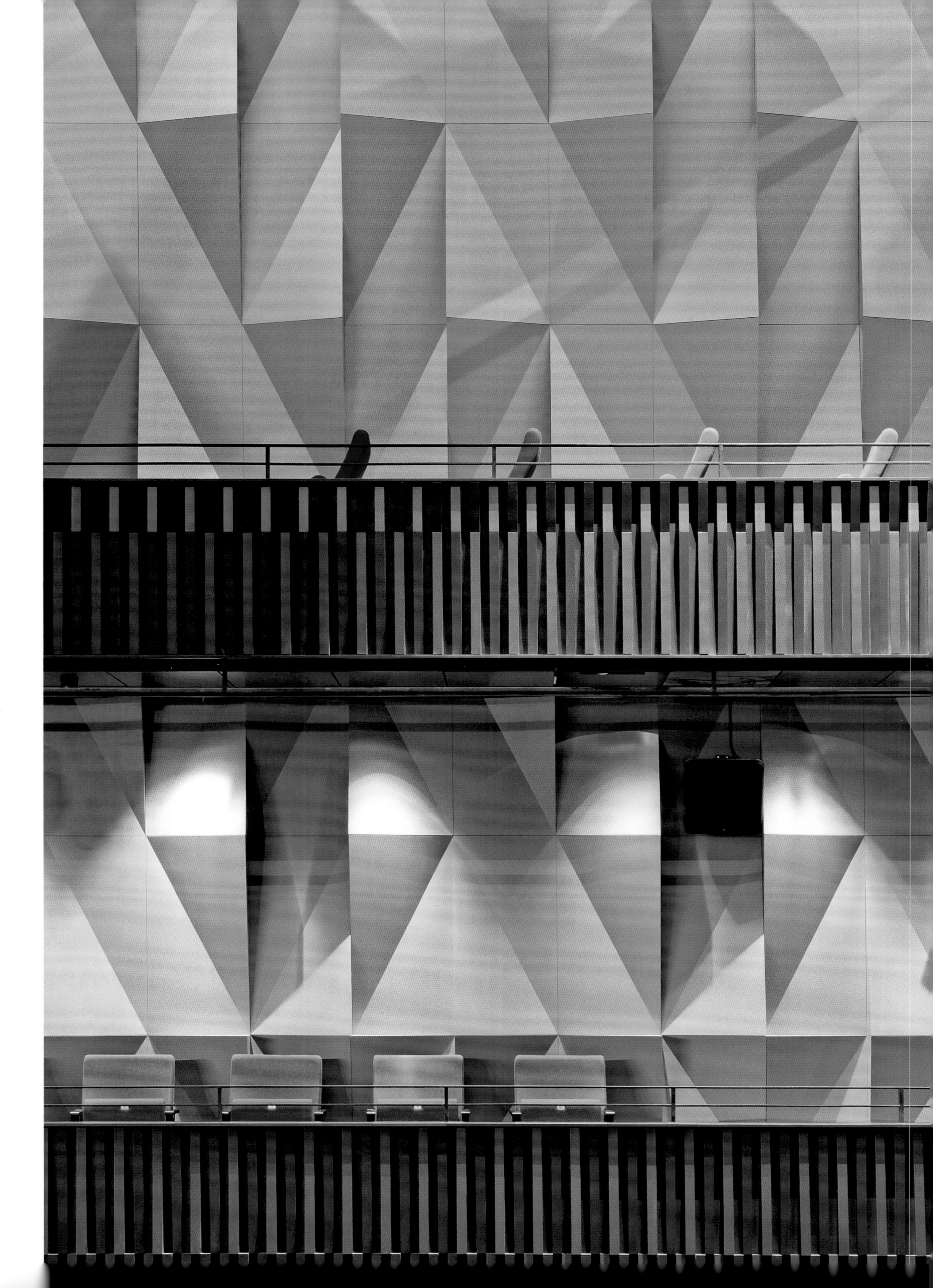

嵌入场地的建筑

Nestle in

在地面嵌入一座建筑物需要对场地具有高度敏感性。这种高度敏感性可以创建一种极为内向的空间，保护其不受恶劣天气影响或遮挡住路人的目光，还可以带来一些显著的环境效益。当建筑立面的大部分都位于地下时，建筑师如何确保建筑仍然可以进行足够的照明和通风？建筑的哪些部分可以呈现在外呢？如果建造的是一座公共建筑，怎么保证这座密封的半地下建筑与外界之间的联系？本文所列举的项目对此采取不同的角度：从奥钢联精神关怀办公室的珠宝掩埋式手法，到主导景观从而自然地创造出舒适环境条件的日间社区中心，再到围绕景观布局的毛伊岛悬崖上的房子。

Embedding a building in the land demands a heightened sensitivity to the site. It can create an intensely introspective space, provide shelter from the elements or from the gaze of passers-by and bring some significant environmental benefits too. But how do you ensure that a building is still adequately lighted and ventilated, when many of the building's facades are built up within earth? Which views do you allow out? If the building is a public one, how do you ensure engagement between a hermetic, semi-underground building and the outside world? The projects presented here take a different perspective on this – from the buried jewel of Pastoral Care Voestalpine, to Day and Community Centre which manipulates its landscape to naturally create comfortable environmental conditions, to Clifftop House Maui that is organized around views.

挖掘掩埋
Digging in

奥钢联精神关怀办公室应建设在地下，即使建在地下要牺牲教堂所需要的外观可见度，这是很恰当的。如建筑师所指出的，传统的解决方案在任何情况下都不再适用。受交通繁忙的道路、当地的钢铁和采矿业的超大型建筑的限制，带有尖顶的独立教堂从此不复存在。

由X Architekten设计的、位于奥地利林茨的奥钢联精神关怀办公室是低调的，从路面上只能勉强看见。作为当地社区的重要建筑，它具有异常温和的形式。与周边建筑不同，它在场地中是下沉的。事实上，这座建筑和本文展示的其他两个项目都表现出：下沉的建筑有自己的力量，它们创造了一个看起来极为内向的空间，保护其不受恶劣天气影响或遮挡住路人的目光，还可以带来一些显著的环境效益。

该建筑项目还要求对场地具有高度敏感性。建筑师需要选择隐藏这类建筑物的部位以及向外界开放的部位。在带有负面因素（大路或重工业区的噪音、强风）的环境中，人们迫切地想保护建筑不受其影响，有些场地可能明显适于建造遮护景观，但在其他情况下这一战略可能不太可行。当建筑立面的大部分位于地下时，建筑师如何确保建筑仍然可以进行足够的照明和通风？建筑的哪些部分可以呈现在外呢？如果建造的是一座公共建筑，怎么保证这座密封的半地下建筑与外界之间的联系？

对土壤进行开采也可以为设计师提供额外的廉价材料。建造建筑物的基础需要开挖泥土，这些泥土通常需要从场地上运走，成本往往较高，因此，将它重新用作景观设计的一部分具有非常重要的意义。周围景观的地形可以根据对建筑物有利的方面进行巧妙处理，以创造私密空间，或创造出用户专用的室外封闭区域。

对于奥钢联精神关怀办公室而言这仅是出发点。它受高高的路堤保护，不受道路影响，游客经由一条通向下沉式庭院和洞穴式入口的切口进入建筑物。无论是从其低矮、笨重的形式还是从其防御性方式（两侧延伸进斜坡，其他立面带有2m高的玻璃带，更大程度地向外界开放）

It is quite fitting that the Pastoral Care Voestalpine should be buried in the ground, even if that comes at the expense of the iconic visibility usually demanded of churches. As the architects pointed out, the traditional solution wasn't appropriate in any case. Bounded by heavy traffic roads and the monumental-scale buildings of the local steel and mining industry, a free standing church with spire would have been lost.

Unassuming and only barely visible from the road, the Pastoral Care Voestalpine in Linz Austria by X Architekten has an unusually modest form for such an important building in the local community. Unlike the neighboring buildings, it hunkers down on its site. In fact, as this and two other projects featured here show, sunken buildings have their own power, creating an intensely inward looking space, providing shelter from the elements or from the gaze of passers-by and bringing some significant environmental benefits, too.

It also demands a heightened sensitivity to the site. Choices need to be made about where to conceal the building and where to open it up to the outside world. In an environment with negative factors from which you want to protect the building – noise from a large road or heavy industrial area, strong wind – there may be obvious places in which to build up a sheltering landscape, but in other instances, the strategy is less clear. How do you ensure that a building is still adequately lighted and ventilated, when many of the building's facades are built up within earth? Which views do you allow out? If the building is a public one, how do you ensure engagement between a hermetic, semi-underground building and the outside world?

Working with the soil can also offer the designer an extra material to work with, and which is, a cheap one. The earth that is excavated in order to construct a building's foundations usually has to be removed from the site, often at high cost, so that its reuse as part of the landscape design makes a huge amount of sense. The topography of the surrounding landscape can be manipulated to the building's advantage, to create privacy, or to create enclosed outdoor areas for the exclusive use of the users.

This is the starting point for the Pastoral Care Voestalpine. Protected from the road by high earth berms, visitors approach the building via a cutting which leads to a sunken courtyard and cave-like entrance. The building is decidedly bunker-like, both in its low, hulking form and the defensive way in which it is built into the slope on two sides, with the other facades opening up to the outside to a greater extent with a 2m high strip of glazing. True to the site, both the banks and Pastoral Care Voestalpine itself have been constructed from the local mining industry's excavation waste – a dark stone that is reconstituted as pebble dash to clad the build-

奥钢联精神关怀办公室，沉入场地
Pastoral Care Voestalpine, hunkers down in its site

上看，它都很像碉堡。路堤和奥钢联精神关怀办公室忠于现场，利用当地采矿业开挖的废物建成，这种废物是一种暗色的石头，被处理成小卵石，以覆盖建筑外墙。

石材和混凝土的灰色延续到室内，至少入口和更为私人的工作空间，如办公室和会议室，都是灰色的。建筑进深大意味着日光需要被引入项目的中心，而这通过创建一个长方形的庭院得以实现。此空间内还设有教堂的钟，又一次参考了该地区的采矿遗产。钟的正下方有一个浅坑，可以像矿井一样将声音扩大，并使声音传播出去。

在灰色的天空下，无色的极简艺术营造了一种阴郁气氛。与此相反，该中心其余部分的清冷空间是酒吧、活动室和小教堂等公共场所。这些场所使用了大量朴素的暗石，显得不拘一格。这些空间的墙壁和天花板具有不均匀的、水晶般的几何形状，由围绕房间的条状三角面组成。一排细长的荧光灯管嵌入面板中，与条状图案吻合，并提供照明。

将目标空间按照岩石内镶嵌宝石的想法进行设计是十分合适的，但很可惜的是，建筑师并没有采取进一步行动。小教堂沿着两个未掩埋立面中的其中一个分布，而建筑理念似乎要求这个空间在建筑中掩埋得更深，以成为一个待发掘的场所。小教堂的体量即使再削减一点看起来也仍然很壮观，而视觉上少一点哗众取宠更能增强其作为建筑礼拜中心的力量。

西班牙马略卡岛帕尔马的日间社区中心是由Flexo Arquitectura设计的，它与类似的掩埋建筑有着更有趣的联系。很显然，该地的气候使得建筑氛围更轻快，灿烂的阳光和较浅的色调减少了阴沉气氛。对于阳光和风的防护是建筑设计的重要因素，同时室外需要有更大的开放性，以应对地中海这一地点对气候的要求。

该建筑掩埋入场地的方式非常与众不同，它利用自然地形和人造结构相结合的方法来划分各个领域。该建筑排列在一条直线上，围绕场地的三面，在中心建立一个私人庭院。该建筑物与周围的街道完全隔绝，为中心营造了一个平静的绿洲。该建筑对外部公共空间的态度是高傲

ing's exterior.

The grey of stone and concrete continues once inside, at least within the entrance and for the more private work spaces such as the offices and meeting rooms. The depth of the building means that daylight needs to be introduced to the centre of the plan and this is achieved through the creation of a rectangular courtyard. This space also houses the church bell, with another reference to the area's mining heritage. A shallow pit has been created directly beneath the bell to amplify its sound in the same way as a mine shaft would, allowing the noise to carry.

Under the grey sky, the colorless minimalism makes for a gloomy atmosphere. In contrast, amongst the sober spaces of the rest of the centre are public programs of the bar and event space, and the chapel. After so many unornamented dark stones, these spaces are riotous. The walls and ceilings there have an uneven, crystalline geometry, made of striped triangular surfaces which fold around the rooms. An array of narrow fluorescent tubes set into the panels, and following the striped patterns, provide the lighting.

This idea of designing the destination space as a jewel set within rock is a fitting one, but it is a pity that the architects did not take it further. The chapel is located along one of the two unburied facades, when the concept would seem to demand that this space be buried deeper within the building as a place to be discovered. It could also have been spectacular with the volume turned down a little – less visual noise would have ramped up its power as the liturgical heart of the building.

The Day and Community Centre in Palma de Mallorca, Spain, by Flexo Arquitectura brings a more playful touch to a similarly buried building. It is clear that climate plays a role in the lighter mood, with bright sun and a lighter color palette providing a less sombre atmosphere. Shelters from the sun and wind are important factors in the design of the building, as is the need for a greater openness to the outside, in order to deal with the climactic demands of the Mediterranean location.

This building takes a very different approach to submerging itself in the site, using a combination of natural topography and man made construction to demarcate the various areas of the plot. The building is arranged in a line, running around three sides of the site to create a private courtyard at the centre. Completely blocked off from the streets around, it creates a calm oasis for the centre. The stand-offish, defensive attitude to the public space

日间社区中心，隐入场地内，在中央创造出一个私人庭院
Day and Community Centre, submerging itself in the site and creating a private courtyard at the centre

的、防御式的，这是由建筑物的半下沉状态决定的（也应该说，是由窗户上的金属格栅决定的），形成了一个集中内省的中心。

如果该中心仍然成功地保留了开放性，它很可能是考虑到了气候因素，气候要求建筑的立面需要分层和遮阳，还需要充足的通风。炎热的夏天，阳光直射到玻璃上增加的热量需要保持在最低限度，因此，庭院周围的大型悬挑结构因为使用竹子而变得轻盈，在下面形成了不均匀的阴影图案。种植园坐落在比建筑本身还要高的地方，它不像硬质景观那样吸收太阳的热量，这也有助于保持该场地的凉爽。反过来，这种凉爽的空气有助于建筑通风。楼层平面的进深较小，而且建筑物的两侧都有窗户，意味着建筑很容易进行被动通风和降温。

场地和景观在建筑物气候调节方面发挥了一部分作用。庭院朝南，因此能得到充足的阳光，但它同样可以保证获得的阳光不过量。夏天中午太阳高度角较大，这意味着阳光很容易被遮阳设备挡住，或像该项目一样，被布局合理的树木挡住。冬天的太阳高度角低，有可能足够低到遮阳设备以下，但在冬天，多余的热量一般是很受欢迎的。

房间是顺着庭院周围的走廊布置的。从某些方面来看，为交通流线空间保留最佳景观并不好，但它实际上意味着这个区域变成了一个公共空间，可以作为有遮阳设施的天井的延伸结构使用，并为室内外之间提供更微妙的连接机会。景观在这里也起着作用，随着抬升的花园挡土墙蜿蜒进出立面，而形成了天井区。

在一些地方，建筑物的外墙成为周围地块的挡土墙。这为建筑物提供了蓄热体，有助于夏季保持凉爽，冬季保持温暖。土地升温慢，冷却得也慢，因此它可以作为保温层，同时通过储存热量和冷气（可以慢慢地渗透回建筑），保护建筑不会达到极端温度。当然还有其他获得蓄热体的方式，如通过在沉重的砖石或混凝土上建造部分建筑物，此外，建筑物周围的景观也可以起到类似的作用。

建筑物内部像白昼一样明亮。涂刷成白色的墙壁和高度抛光的地板表面自然有助于实现这一情形，但马略卡岛的太阳光线的作用也不能

outside, established by the decision to half-sink the building (and also, it should be said, the metal grills covering the windows), creates an intense introspective focus.

If the centre still manages to retain a sense of openness it is probably down to the climate, which creates a need for layering and shading in the facade and for ample ventilation. Hot summers mean that the heat gains caused by direct sun on glass need to be kept to a minimum, hence the large overhangs around the courtyard lend a playful lightness through the use of bamboo to create an uneven pattern of shadow below. The planted garden, sitting at a higher level than the building itself, doesn't absorb the sun's heat the way that hard landscape would, which also helps to keep the site cool. In turn, this cool air can help ventilate the building. The shallow depth of the floor plan and the fact that there are windows on both sides of the building mean that it is easy to ventilate and cool the building passively.

The site and the landscape also play their part in regulating the building's climate. The courtyard is orientated south, so that it gets plenty of sun, but it is equally possible to ensure that it doesn't get too much. The high angle of the sun at midday in the summer means that it is easily blocked out by shading devices, or as is also the case here, by well placed trees. The lower winter sun might sink low enough to reach under the shading devices, but at that time of year the extra heat is generally welcome.

The rooms are organized along a corridor which follows the courtyard. In some ways it seems a shame to reserve the best view for the circulation space, but what it actually means is that this area instead becomes a public space, to be used as an extension of the shaded patio and offering the chance of a more subtle connection between inside and outside. The landscape also plays a role here, shaping the patio area as the retaining walls of the raised garden zigzag in and out from the facade.

In several places the exterior walls of the building are actually retaining walls for the surrounding earth. This provides thermal mass for the building, helping to keep it cool in summer and warm in winter. Earth both heats up and cools down slowly, so it can protect the building from extremes of temperature by acting as insulation, as well as by storing heat and "cool" which can be slowly released back into the building. There are other ways to achieve this thermal mass of course, by building parts of a building in heavy masonry or concrete, but the landscape surrounding a building can work, too.

完全被忽视。这个项目提出了这样一个问题：如何尽可能成功地掩埋建筑物，同时还能保证高纬度国家充足的自然光照明。在这里，阴影是很受人们欢迎的，即使屋顶的屋檐悬挑很大，也仍然可以实现高质量照明。再往北，如果建筑物仍然需要很好的自然照明，就需要设计师在设计时对建筑物进行仔细思考。

洞口如何设计可以让光线和空气进入，并让人们看到外界景色是Dekleva Gregoric Arhitekti设计的夏威夷毛伊岛悬崖上的房子的主要布局考虑因素。该场地较孤立，位于俯瞰大海的悬崖上。这座建筑的私密性需求相对于遮蔽性需求要低。它不是沉入场地中，而是将景观向上拉起，使其高度超过自身。峰巅和屋顶的浅梯度反映了岛上的火山地貌。夏威夷群岛的黏稠熔岩意味着喷发导致了相对平缓山坡的形成。房子的屋顶被设计为周围环境的延续结构，意在调整规模，供人们行走，成为业主的另一处户外空间。

所有的材料都源自本地。屋顶采用的本地材料是木材。木材在飞檐下（此处是多加庇护的地方）保留了其天然红色，上面曝露在风吹日晒中，已经风化成银灰色。屋顶木板呈条状，顺着屋顶的斜坡安放，而不是横向放置，屋顶的外壳看起来似乎是从组成小岛的火山岩层里雕刻出来的。

从崖顶地区可以俯瞰到大海的壮观景色，建筑物的几何形状经过了精心安排，以充分欣赏海景。屋脊呈扇形散开，在其下形成了一系列空间，这些空间是对外开放的。从那里可以看出，观景视野的设计被房屋内一系列棋盘状排列的封闭箱体结构代替。它包含不同的空间：一个车库和为客人准备的卧室。两者之间剩余的空间包含房屋更为公共的职能，特别是厨房和用餐区。

屋顶的形状，连同房子墙体的分散模式，意味着所有房间（无论封闭式还是开放式）的视野都被强有力地引导至外界的景观。该建筑的相对裸露性在某些方式上与其他两座建筑物形成鲜明对比，但屋顶仍成功地创造了防护性封闭领域。露台恰好位于建筑物的周围，具有同样的品质，既对外开放又提供遮护，木质立面的红色即是有力证明。

正如在马略卡岛，建筑物的屋顶具有遮阳的功能，尤其针对西南

Inside the building appears brightly lit as day. White painted walls and the highly polished surfaces of the floors naturally contribute to this, but the strength of the sun light in Mallorca cannot be entirely discounted. It raises the question of how successfully it is possible to submerge buildings and thereby still get the chances of adequately lighting them naturally, in countries at higher latitudes. Here, shade is welcome and deeply extended eaves of the roof still allow a good quality of lighting; further north, care is required if a building is still to be naturally well lit.

The question of how openings can be created to allow light and air in and views out, is the main organizing factor in the Clifftop House Maui, Hawaii by Dekleva Gregoric Arhitekti. The location is a reasonably isolated one, on a cliff looking out over the sea. There is less need for privacy than for shelter. Rather than sinking itself down into the site, it actually appears to pull the landscape up and over itself. The peaks and shallow gradients of the roof reflect the island's volcanic topography. The viscous lava of the Hawaiian islands means that eruptions result in the creation of relatively gentle slopes. The house's roof is designed as a continuation of these surroundings and is intended to be scaled and walked upon, as a further outdoor space for the residents.

All of the materials were sourced locally. In the case of the roof, this means wood. It retains its natural red under the overhanging eaves, where it is more sheltered, weathering to silver grey on top where it is exposed to sun and wind. Lain in strips which run in line with the roof's slope, rather than cutting across it, the roof's shell appears to have been carved out of the layers of volcanic rock which make up the island.

From the clifftop site there are spectacular views out over the sea and the geometry of the building has been carefully arranged to make the best of them. The ridges of the roof fan out to create spaces below which open up towards the exterior. From there, the game of setting out views is taken over by a series of closed boxes arranged through the house in a checkerboard-like pattern. They contain variously, a garage, and bedrooms for guests. The leftover space which flows between them contains the more public functions of the house, notably the kitchen and dining area.

The shape of the roof, together with the spreading pattern of walls in the house, means that all the rooms are forcefully directed towards the view, closed and open rooms alike. The building's relative exposure is in some ways in sharp contrast with the other two buildings, but the roof still manages to create areas of protective enclosure. The terraces which are situated right around the building have this quality, at once open to the air and sheltered, as evidenced by the redness of the wooden facade.

As in Mallorca, the roof serves a shading function, particularly to

毛伊岛悬崖上的房子，不是沉入场地中，而是将景观向上拉起，使其高度超过自身
Clifftop House Maui, rather than sinking itself down into the site, it actually appears to pull the landscape up and over itself

部，屋顶用于保护户外露台和房屋玻璃免受强烈阳光的直射。自然通风能够使房子凉爽。由于遮阳首先就防止了过热，因此这就足以控制室内温度，且不需要使用空调。

如果以这种方式可以对建筑产生显著的可持续发展效益，也许我们还需要做到最后一点以达到某种平衡。这些建筑对土地具有令人难以置信的狂热。如果可以对场地进行灵敏地回应，它们几乎可以完全消耗这些土地。Atelier Carvalho Araújo建造的Mário Sequeira美术馆位于葡萄牙布拉加。这座建筑物半埋进山坡，朴素的白色建筑露出地面，使人们几乎看不到建筑物内部发生的事情。建筑师说，与外界隔绝的感觉是为了提高游客走近大楼时的期待感。这种地下的位置也适合建造一间艺术画廊，在这里"封闭的盒子"意味着可对环境进行精确控制，这对任何敏感的艺术品都是有利的。该美术馆仍然成功地与外部相连，裸露的东侧立面的大窗户可使许多景色尽收眼底。

意大利—委内瑞拉中心的新服务大楼是由Roberto Puchetti设计的，在一个大屋顶下形成了很多空间。这些空间下面的立面布局设计回应了屋顶形成的阴影图案，以避免阳光直射过热而形成的不适宜温度。这一策略使得楼内不需要安装空调，而在这个国家，空调的使用很普遍。屋顶也能够收集到足够的雨水，可以对厕所进行冲刷和清洗建筑物内部，也可以冲洗运动场地。

西班牙潘普洛纳某托儿所由Pereda Pérez Arquitectos设计，嵌入了一处人造景观内。它占据了一个三角形场地，这个场地的其中两边是由原有建筑物的封闭性后墙组成的。为了给运动场留下一个大的、阳光明媚的空间，建筑师选择将学校建造在场地的北部边缘，建造一座进深较大的建筑，只有一个立面暴露在外。教育区域建造在这里，以与运动场产生直接联系，而管理和设备空间设在后面。为了保证这些不同寻常的空间有良好的照明和通风，建筑师已经开发出一种采光天窗系统和庭院，这使得大型机构建筑具有小建筑的私密感。南部的立面具有类似的功能，混凝土梁和柱避免了阳光直射，提供阴凉处的同时也创造了一个更私密的、占据整排教室长度的柱廊。

the south west, where it protects both the outdoor terraces and the house's glazing from strong direct sunlight. Natural ventilation cools the house. Since shading prevents it from overheating in the first place, this is sufficient for controlling the indoor temperature and air conditioning has been done away with.

If there are significant sustainability benefits to building in this way, perhaps one last point needs to be made in counter balance. These buildings are incredibly land hungry. If they may respond sensitively to their sites, they can consume them almost entirely.

Mário Sequeira Gallery by Atelier Carvalho Araújo is located in Braga, Portugal. Half-buried in the hillside, the plain white volumes which emerge from the ground give away little about what is happening inside. This feeling of being closed to the outside is to heighten visitors' sense of anticipation as they approach the building, say the architects. Such an underground location is also appropriate to an art gallery, where the "closed box" means that the environment can be carefully controlled, to the benefit of any sensitive artworks. The gallery still manages to make a connection to the exterior, however, with large windows on the exposed eastern facade allowing views out.

The Italian-Venezuelan Centre New Services Building by Roberto Puchetti unites a number of spaces under one large roof. The layout of the facades below it was designed in response to the pattern of shadow that the roof creates, to avoid direct sunlight from entering the building and heating it to an uncomfortable temperature. This strategy has made air conditioning unnecessary, in a country where it is prevalent. The roof is also able to capture enough rain water for flushing the toilets and cleaning within the building, as well as to irrigate the sports fields.

This Nursery School in Pamplona, Spain by Pereda Pérez Arquitectos is embedded in a man-made landscape. It occupies a triangular site, where two sides are bounded by the closed rear walls of existing buildings. To leave a large sunny space for the playground, the architects choose to build the school up against the northern edges of the site, creating a deep building, with only one exposed facade. The education programme is built here, so that it has a direct connection to the playground, while the administration and service spaces are behind. To ensure good lighting and ventilation to these out-of-the-way spaces, the architects have developed a system of roof lights and courtyards, which bring a small scale intimacy to a large institutional building. The southern facade serves a similar function, with the concrete beams and columns providing shade from direct sunlight, while also creating a more private colonnade which runs the length of the row of classrooms. Alison Killing

绿洲——奥钢联精神关怀办公室

X Architekten

林茨主教区的精神关怀办公室将建在奥钢联钢铁厂的场地上，既可用来做礼拜仪式，也有实际用途。这个地块位于主干道和工业区之间的无人区，所以建筑本身需要鲜明的特色。

建筑师对钢铁厂内的人和工作进行了研究，从而形成了现在的理念——通过"建筑景观"在建筑基地内补充建造。现存的绿色林地成为该建筑理念的起点——一片"人们的绿洲"，而且新的建筑景观因为嵌入在山坡中也保留了绿洲这一特征。传统意义上的教堂是很难在这个地区成为独立存在的建筑的，因为这个钢厂的场地中主要是相对较高的建筑。

建筑的主干道蜿蜒在山坡里，朝入口处延伸，路面越靠近绿地越宽，最后通向斜坡旁的树林。路堤由黑色的矿渣石构成，矿渣石或者直接铺洒在路堤上，或者加工成小卵石大小。十字架、树林、草地、木棚紧紧围绕在建筑周围。从草地开始，到不同种类的植物，再到松树林，植被逐渐变得浓密。通过走道和楼梯爬上屋顶的途中要经过钢质车棚还有连着公寓和青年活动区的花园。

"建筑景观"的理念也决定了内部的设计。建筑中的"沟壑"将办公室、会议室和工作室这些功能区与包括衣帽间、酒吧、活动室和小教堂的社交和宗教活动区域分隔开。凹进去的"钟室"四角呈直角，散发着一种引人深思的气氛，而且钟直接放置在一层。钟下方的空间可以传播声音（就像矿井里用的铃）。

一个带有白色涂层的木质壳体被分成多个三角形，将小教堂、活动室和酒吧连成一体，成为社交区域的主要空间。这种水晶般的几何形状营造出非凡的氛围，萦绕在既是精神关怀之神又是采矿守护神的圣芭芭拉周围。两面可以滑动的墙体可以创造出适于不同活动的单独或连起来的房间。开放式房间作为一个整体，容纳了小教堂和酒吧，从而创造了一种独特的开放性和完整性。

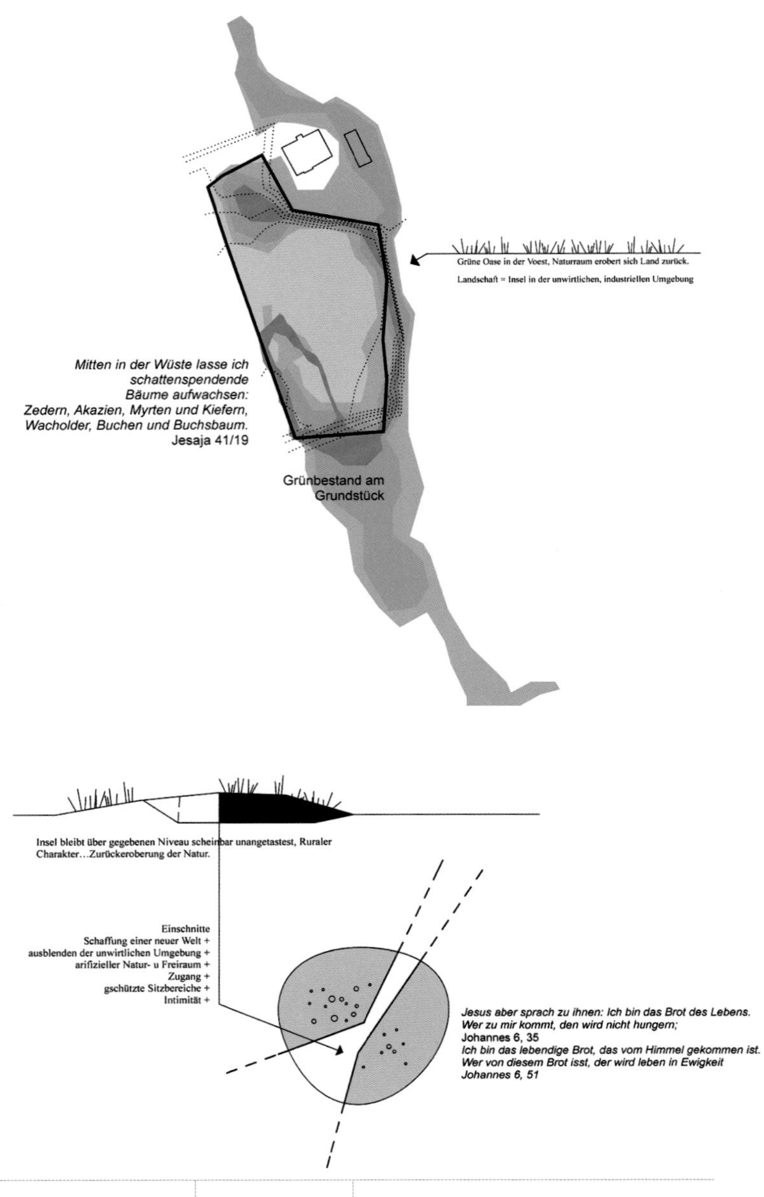

OASIS – Pastoral Care Voestalpine

The office for pastoral care in the diocese of Linz, to be located on the site of the steel company Voestalpine, is to serve liturgical as well as secular purposes. The plot of land lies as a "no-man's-land" between main roads and industrial estates and is in need of a new strong character.

The study of man and work within the steel company led to a complementary addition to the site through its "built landscape". Existing woodlands serve as a conceptional starting point, an "oasis for the people", and the new built landscape retains this char-

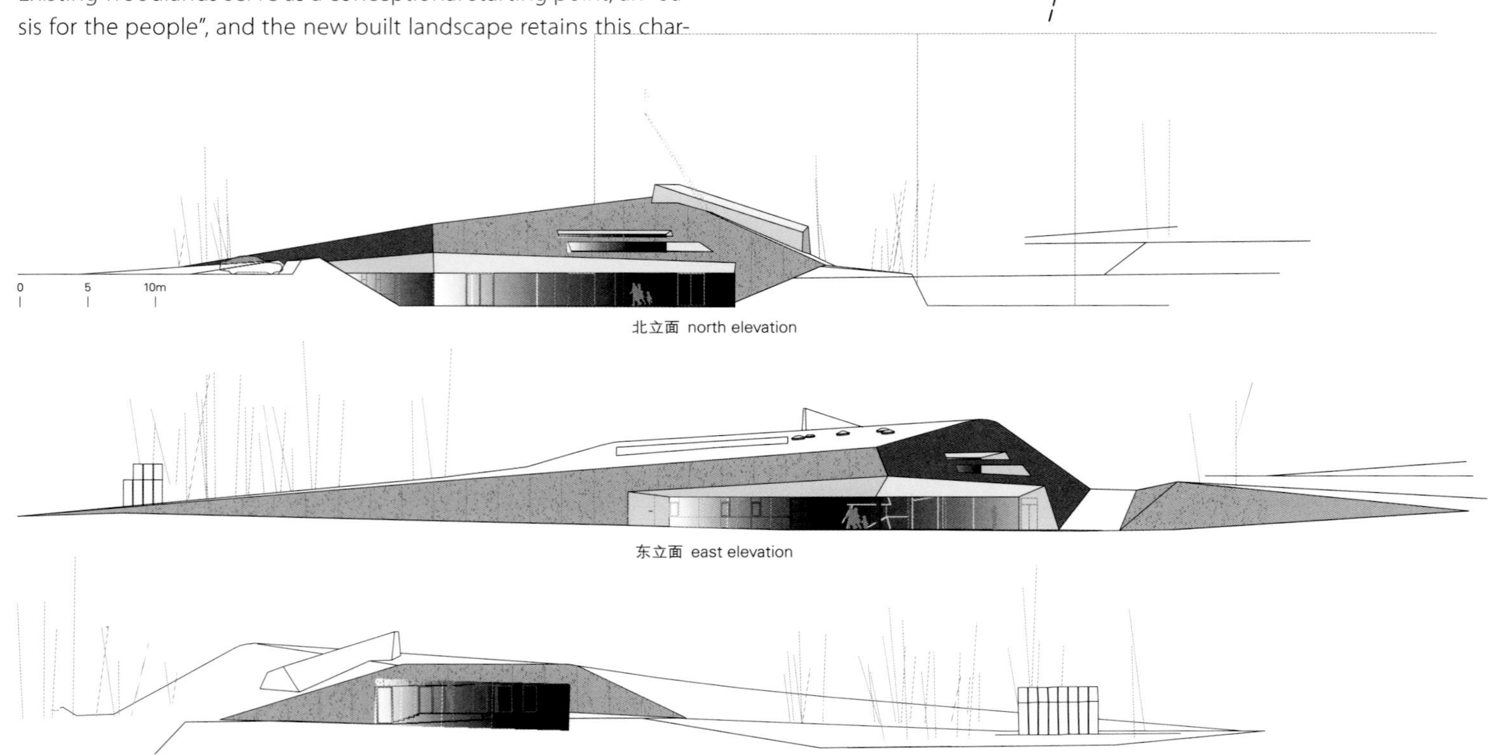

北立面 north elevation

东立面 east elevation

西立面 west elevation

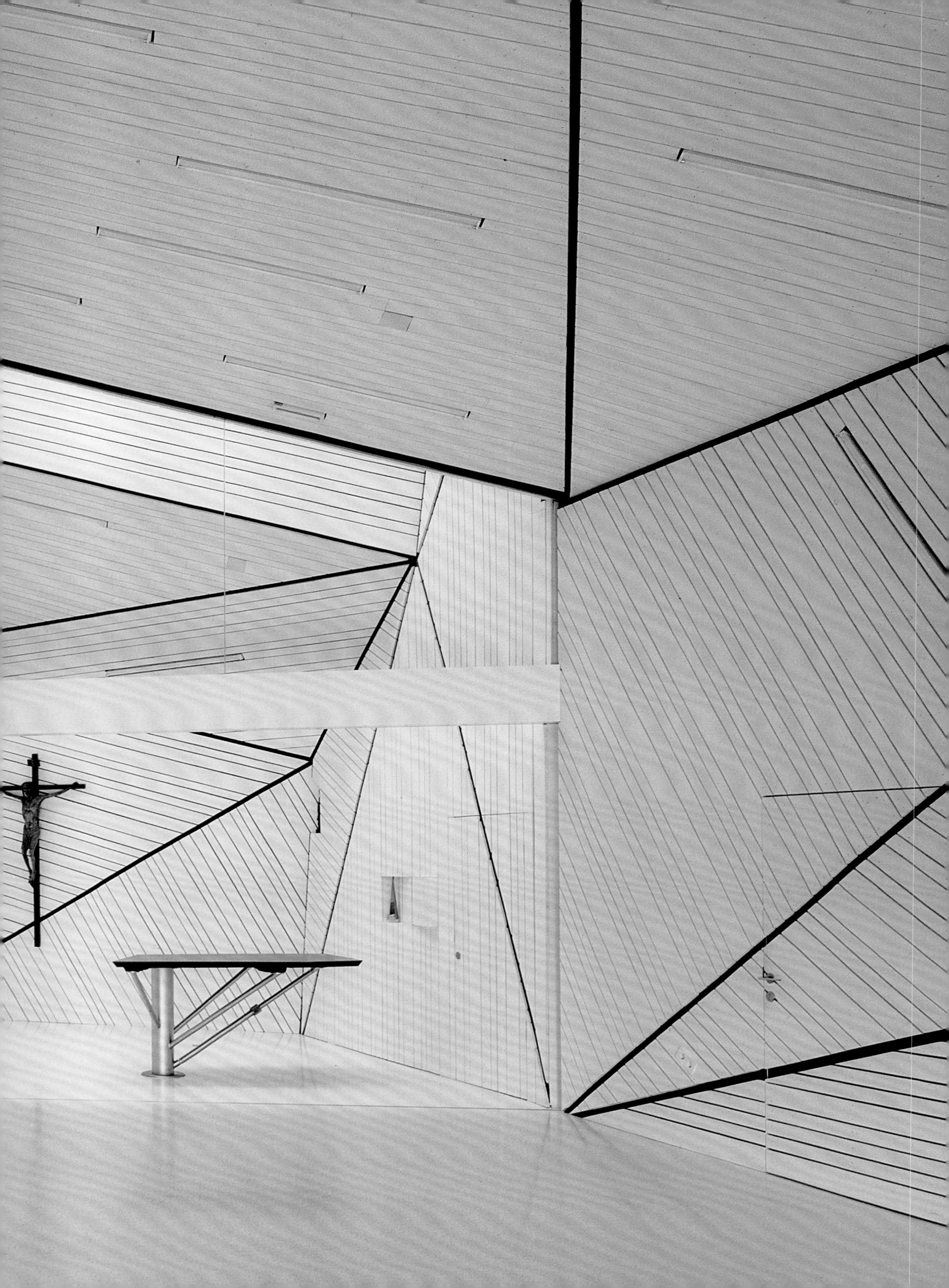

acter by being embedded within a hillside. Conventional church buildings would not be able to achieve their independent meaning in this area as the Voestalpine's site is mainly dominated by relatively high buildings.

The building's main artery, which cuts into the hillside, runs towards the entrances and widens as it approaches the green space, finally leading up to the woods along a sloping surface. The embankment, made of dark slag-stone, is either poured into the embankment or processed as pebble dash. The cross, the edge of the wood and meadow as well as the wooden shed are situated up against the edge of the building. The planting increases in density starting from the lawn, via a diversity of plants, towards the coniferous forest. Climbing the roof via walkways and stairs, one passes the steel car port as well as the garden associated with the apartment and youth area.

The "built landscape" concept also determines the interior. The "ravine" divides the functional areas including offices, meeting room and workshops from the social and religious areas including cloakroom, bar, event room and the chapel. The recessed "bell court" radiates a contemplative mood with its squares and its bell placed at ground-level. The hollow space under the bell serves to distribute the sound (like the bell in the mining shaft).

A wooden and white coated shell divided into triangles unites the chapel, event room and bar as the main rooms of the social area. This crystalline geometry creates an important meaning encompassing Saint Barbara as patron saint of both pastoral care and mining. Two sliding walls enable a choice of separate or connected rooms for different events. The open room as a whole inhabits the chapel and bar and thus establishes a unique openness and integration. X Architekten

项目名称：Oasis – Pastoral Care Voestalpine
地点：Wahringerstraße 30, Linz, Upper Austria
建筑师：X Architekten
甲方：Diocese Linz
用地面积：4,843m² 建筑面积：840m²
设计时间：2008.2 竣工时间：2011
摄影师：©David Schreyer (courtesy of the architect) (except as noted)

照片提供：©Rupert Asanger (courtesy of the architect)

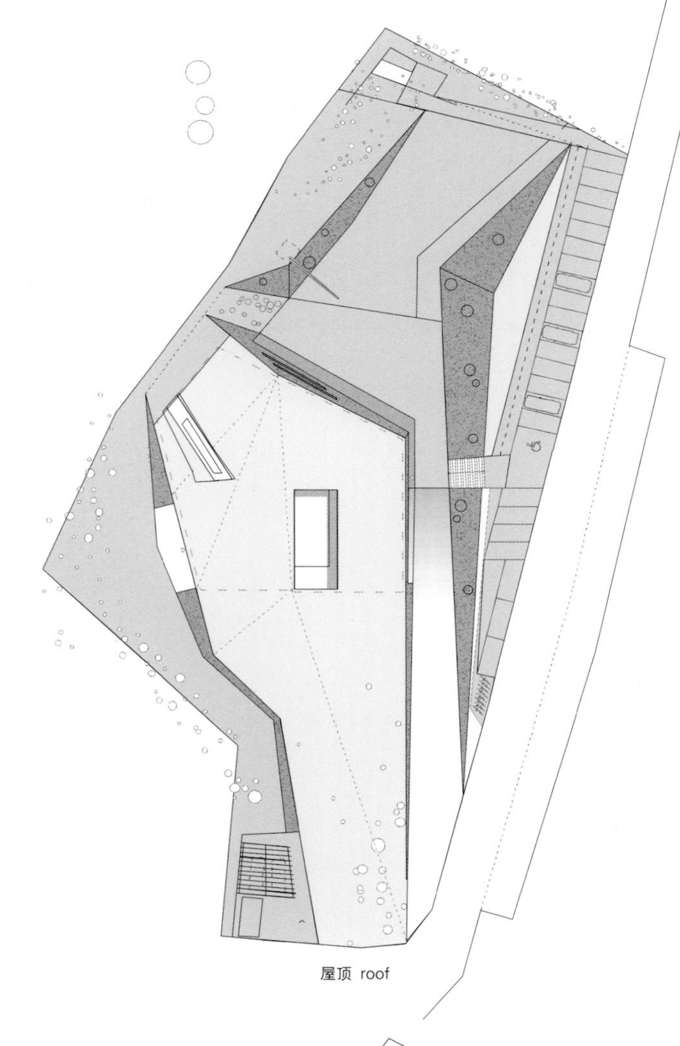

屋顶 roof

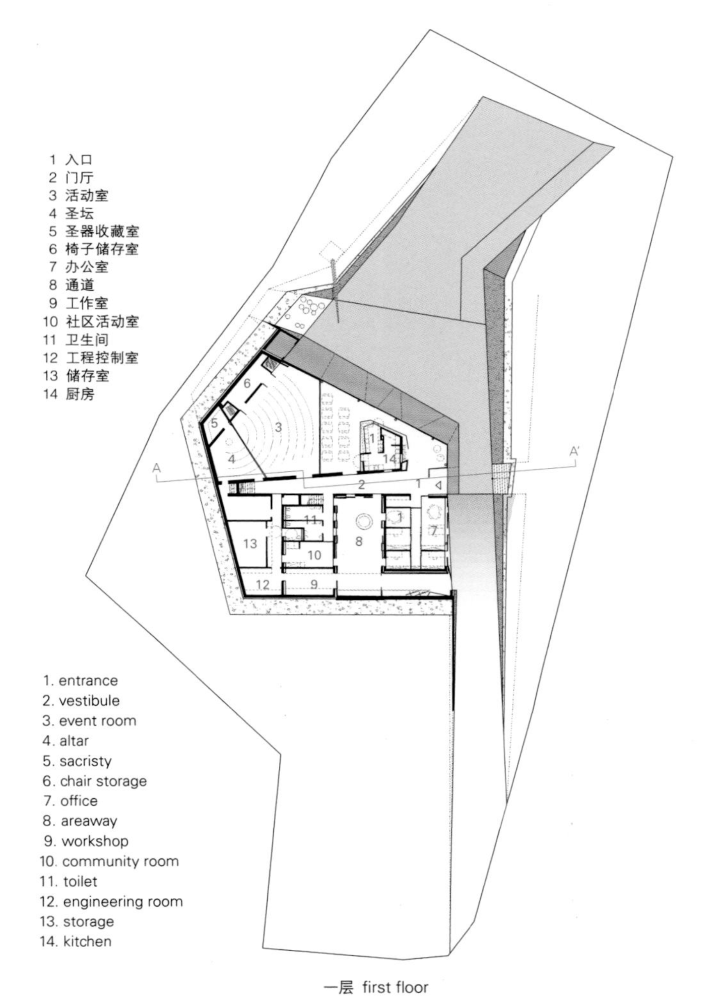

1 入口
2 门厅
3 活动室
4 圣坛
5 圣器收藏室
6 椅子储存室
7 办公室
8 通道
9 工作室
10 社区活动室
11 卫生间
12 工程控制室
13 储存室
14 厨房

1. entrance
2. vestibule
3. event room
4. altar
5. sacristy
6. chair storage
7. office
8. areaway
9. workshop
10. community room
11. toilet
12. engineering room
13. storage
14. kitchen

一层 first floor

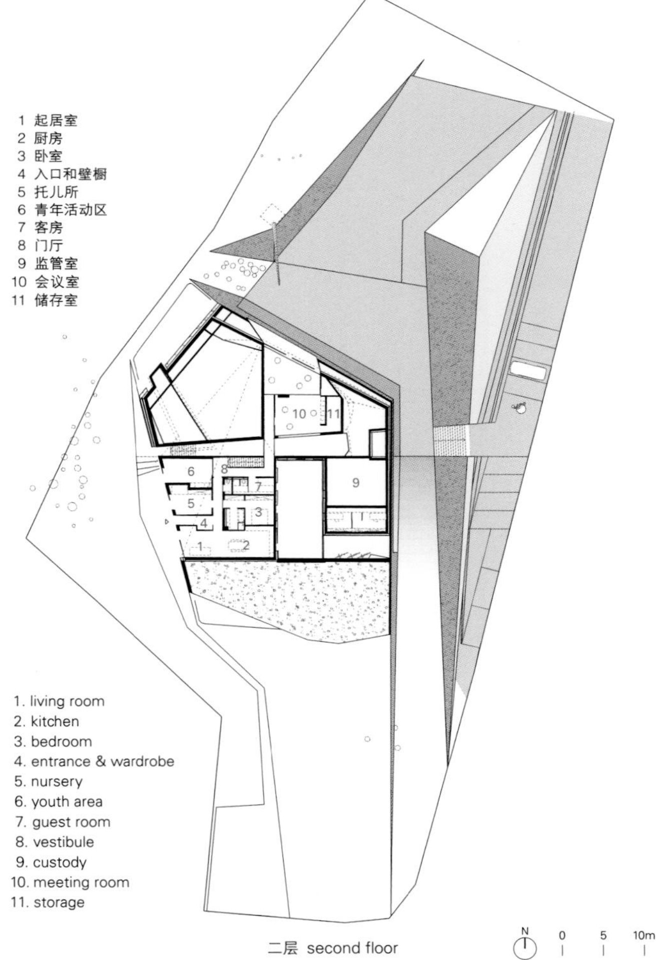

1 起居室
2 厨房
3 卧室
4 入口和壁橱
5 托儿所
6 青年活动区
7 客房
8 门厅
9 监管室
10 会议室
11 储存室

1. living room
2. kitchen
3. bedroom
4. entrance & wardrobe
5. nursery
6. youth area
7. guest room
8. vestibule
9. custody
10. meeting room
11. storage

二层 second floor

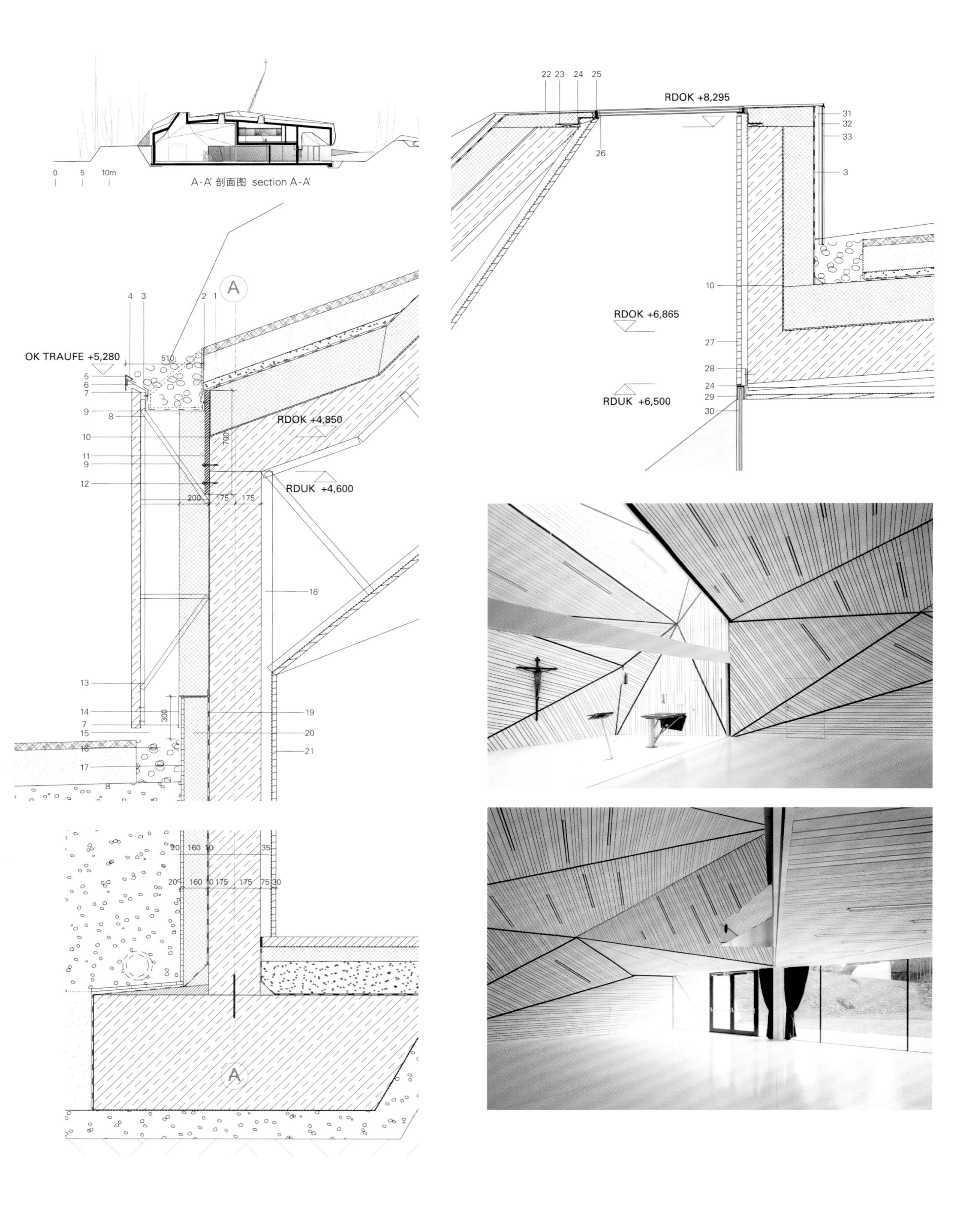

A-A' 剖面图 section A-A'
0
5
10m
OK TRAUFE +5,280
RDOK +4,850
RDUK +4,600
RDOK +8,295
RDOK +6,865
RDUK +6,500

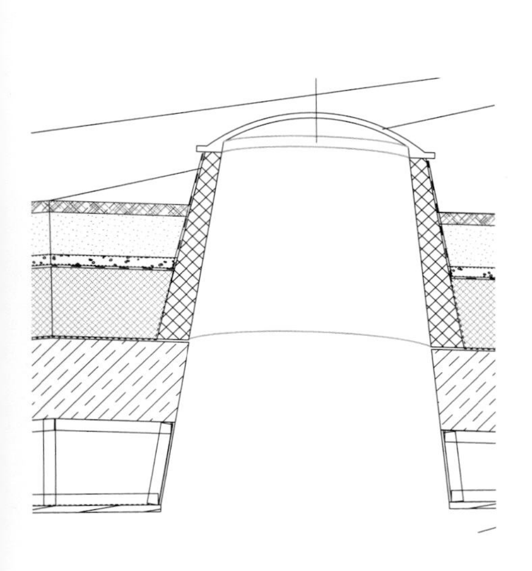

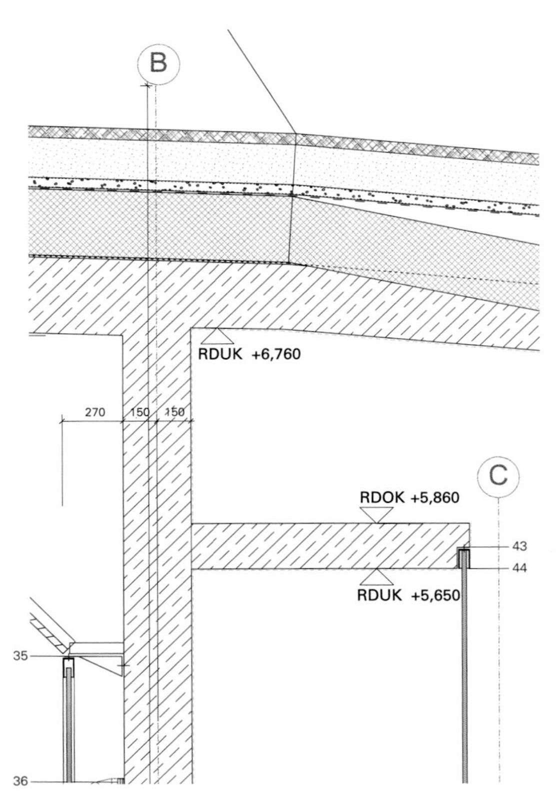

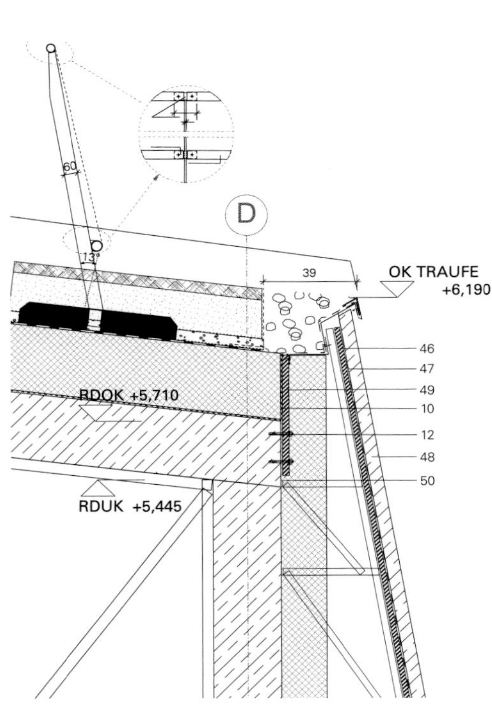

1. hochzug filtervlies
2. winkel lochblech geklebt 280/150
3. hochzug abdichtung
4. verblechung beschichtet innenseiting geklebt
5. folienblech
6. einhängeblech
7. insektenschutzgitter
8. STAKO(noch zu definieren)
9. U-winkel stahl 120/300/80
10. hochzug dampfsperre
11. OSB-platte 3cm
12. klebeanker
13. schlackestein fassadenplatte
14. alu sockelblech 2mm
15. L-winkel 150/150/4
16. kiesstreifen
17. noppendrainmatte 2cm
18. UK für abgehängte decke
19. hochzug feuchtigkeitsabdichtung 50cm über fertigterrain
20. WD-XPS
21. holzschalung gehobelt und gestrichen
22. 1-scheiben glas emailliert
23. Z-winkel 130/40/130/5
24. silikon
25. 3-scheiben verglasung
26. bituthene
27. bretterschalung gehobelt, gestrichen
28. L-winkel 120/60/5
29. U-profil auf stahlwinkel geschraubt
30. fixverglasung 1-scheiben
31. lattung 3/5
32. Z-winkel 2/2/2
33. CORTEN stahl fassadenplatten 1cm
34. lightkuppel DM 140gm
35. holzschalung öffenbar
36. einzelne L-winkel stahl
37. laufwagen zur wartung
38. beleuchtung
39. kirchenfenster
40. einzelne L-winkel stahl
41. abdeckung magnetisch gehalten
42. profil für schiebewand
43. aussparung 10/6cm
44. U-profil auf rohdecke geschraubt
45. fixverglasung 1-scheiben
46. OSB-platte 24mm montage of STAKO
47. abdichtung bitumen
48. sclackestein fassadenplatten
49. OSB-platte 30mm mantage of beton
50. UK-stahl(noch zu definieren)

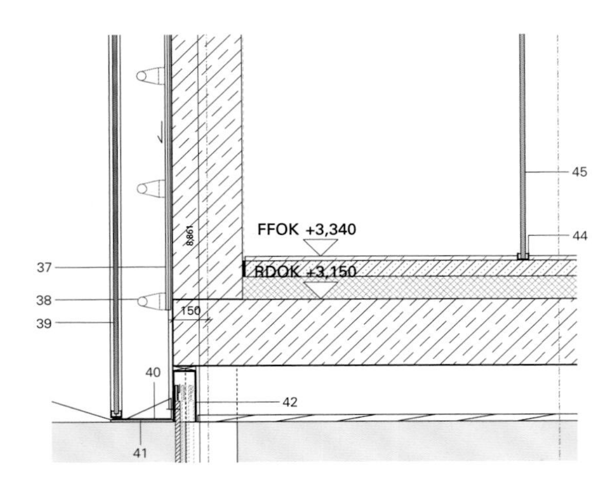

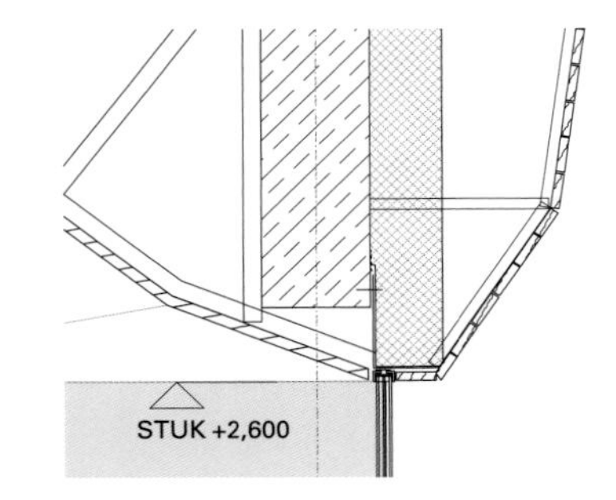

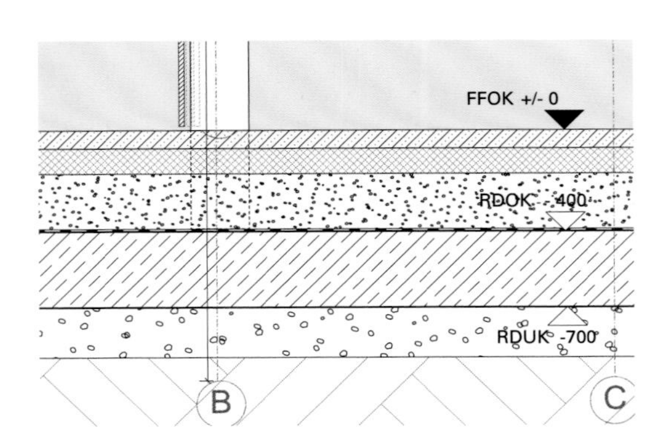

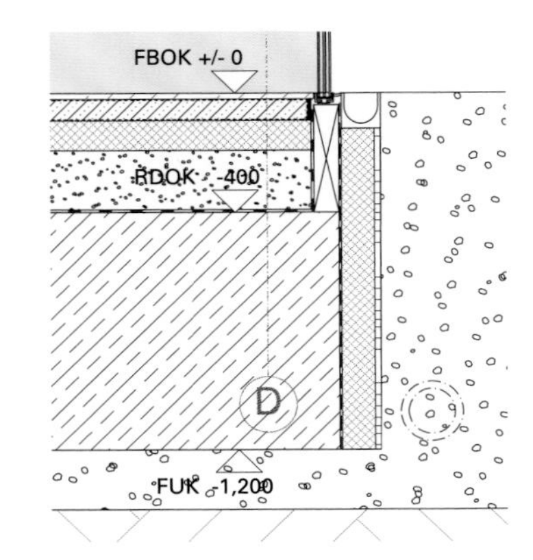

日间社区中心

Flexo Arquitectura

建筑师们被要求在一片郊区居民区（介于帕尔马城市结构和周边区域之间的边缘地带）中建造一座日间社区中心，预算仅为€620/m^2。

他们将该项目视为一次改善该中心人际关系和集体活动的机会。因此，建筑师们将该建筑围绕在一座大型且阳光充沛的私人花园周围，这里能够遮挡住大风，并使行人看不到里面，从而便于在这里举办各种活动。

通过设计其外形，建筑师们决定巩固该中心的机构性质。因此，他们采用了一种与众不同的颜色（白色），旨在使建筑立面上所使用的不同材料具有统一性。

由于经济预算限制在€620/m^2，因此建筑师们提出了一种混合式解决方案，即将当地（工匠）技术和材料与全球（工业）技术和材料结合起来。这样一来，他们可以将二者提供的资源最优化。他们将把建筑形式作为一种可持续性策略，重点强调被动式策略。

总之，带着一种毫无偏见的态度，为任何一个位于特殊文化背景中的项目赋予一层崭新的意义，建筑师对这种能力非常感兴趣，这可以让他们在看似毫无价值的项目中找到机遇。建筑师们提议将特殊的文化背景当作可以树立项目策略的机会。他们选择了一种包含一切且毫无偏见的态度，从而不会忽视这个或那个，取而代之的则是从中选择。

项目名称： Day and Community Centre
地点： C/ Pensament 5. Palma de Mallorca, Spain
建筑师： FLEXO Arquitectura (Tomeu Ramis, Aixa del Rey, Barbara Vich)
合作者： Albert Jener, Cristina Oliver
技术建筑师： Barbara Estudillo
结构工程师： Juan Pablo Rodriguez
机械与电气工程师： Antoni Aguilar
景观建筑师： FLEXO Arquitectura
甲方： Consorci Recursos Sociosanitaris, Conselleria Benestar Social
用途： Community centre
用地面积： 1,440m^2
建筑面积： 530m^2
总楼面面积： 530m^2
造价： EUR 350,000
设计时间： 2009.6—2009.10
施工时间： 2010.2—2010.11
摄影师： ©José Hevia

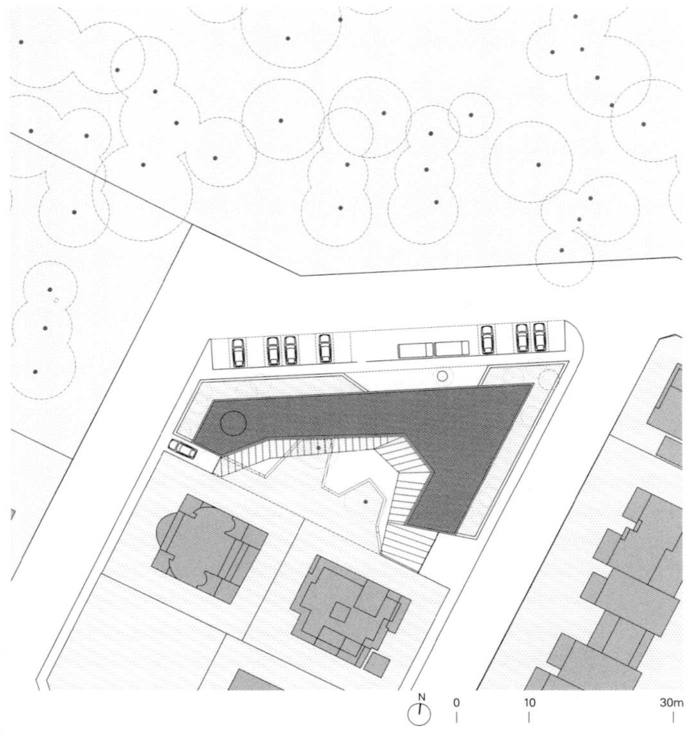

Day and Community Centre

In a suburban residential area, situated at the edge between the periphery and the urban tissue of Palma, architects are asked to propose a day and community centre for a small budget: € 620 / m².

They understand the project as an opportunity to enhance relations and collective activities in the centre. Therefore, the building is arranged around a large and sunny private garden, sheltered from the wind and protected from pedestrians' view, that articulates all the activities.

They propose the consolidation of the centre's institutional character by designing the building's image, therefore, they use a unique colour (white) in order to unify the different materials used on the facade.

Because of the economic budget constraint of € 620 / m², they propose to give a hybrid answer, combining local (artisan) techniques and materials, with global (industrial) techniques and materials. In this way, they can optimise the resources offered by both sectors. Architects will use the architectural form as a strategy for sustainability emphasising passive strategies.

In short, they are interested in the ability of giving any project a new meaning within a specific cultural context with an unprejudiced attitude. It can allow them to find opportunities even in the apparent worthless. They propose using particular cultural contexts as opportunities from which to build the project's strategy. They chose an inclusive and unprejudiced attitude that doesn't ignore this or that but opts for this and that. Flexo Arquitectura

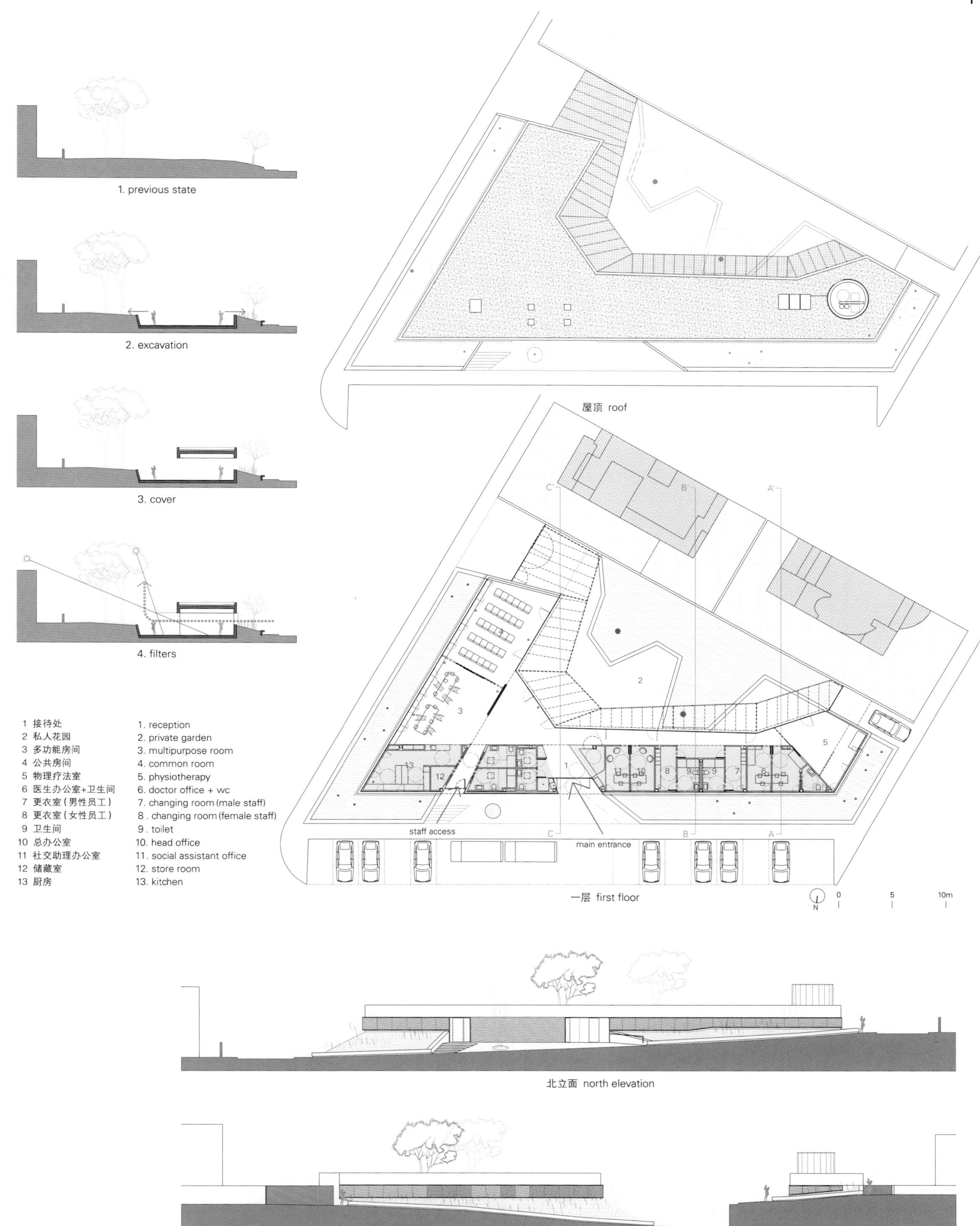
1. previous state
2. excavation
3. cover
4. filters
屋顶 roof
staff access
main entrance
一层 first floor
0
5
10m
N
1 接待处
2 私人花园
3 多功能房间
4 公共房间
5 物理疗法室
6 医生办公室+卫生间
7 更衣室（男性员工）
8 更衣室（女性员工）
9 卫生间
10 总办公室
11 社交助理办公室
12 储藏室
13 厨房
1. reception
2. private garden
3. multipurpose room
4. common room
5. physiotherapy
6. doctor office + wc
7. changing room (male staff)
8 . changing room (female staff)
9 . toilet
10. head office
11. social assistant office
12. store room
13. kitchen
北立面 north elevation
东立面 east elevation
西北立面 north-west elevation

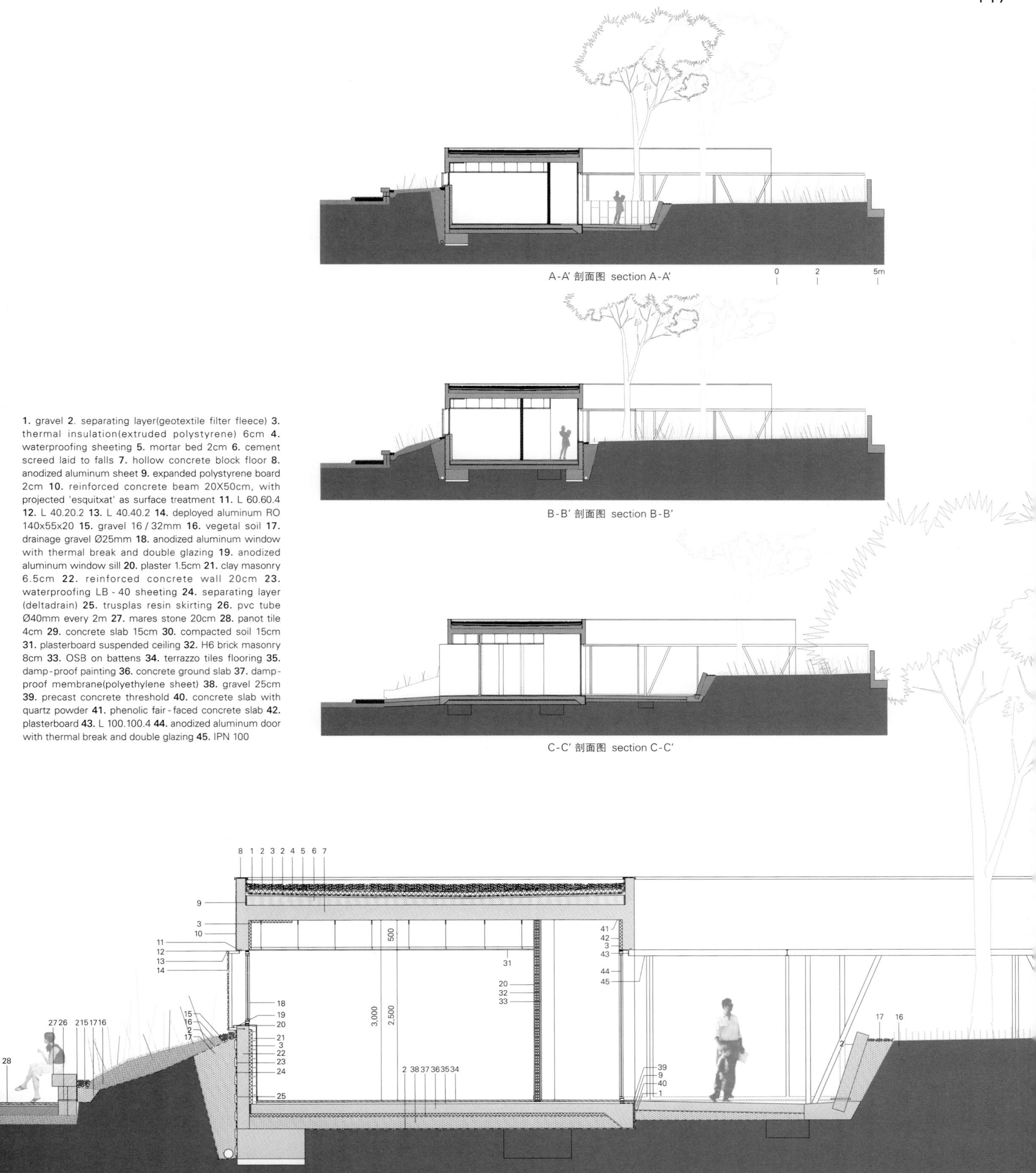

A-A' 剖面图 section A-A'

B-B' 剖面图 section B-B'

C-C' 剖面图 section C-C'

1. gravel **2.** separating layer(geotextile filter fleece) **3.** thermal insulation(extruded polystyrene) 6cm **4.** waterproofing sheeting **5.** mortar bed 2cm **6.** cement screed laid to falls **7.** hollow concrete block floor **8.** anodized aluminum sheet **9.** expanded polystyrene board 2cm **10.** reinforced concrete beam 20X50cm, with projected 'esquitxat' as surface treatment **11.** L 60.60.4 **12.** L 40.20.2 **13.** L 40.40.2 **14.** deployed aluminum RO 140x55x20 **15.** gravel 16 / 32mm **16.** vegetal soil **17.** drainage gravel Ø25mm **18.** anodized aluminum window with thermal break and double glazing **19.** anodized aluminum window sill **20.** plaster 1.5cm **21.** clay masonry 6.5cm **22.** reinforced concrete wall 20cm **23.** waterproofing LB - 40 sheeting **24.** separating layer (deltadrain) **25.** trusplas resin skirting **26.** pvc tube Ø40mm every 2m **27.** mares stone 20cm **28.** panot tile 4cm **29.** concrete slab 15cm **30.** compacted soil 15cm **31.** plasterboard suspended ceiling **32.** H6 brick masonry 8cm **33.** OSB on battens **34.** terrazzo tiles flooring **35.** damp-proof painting **36.** concrete ground slab **37.** damp-proof membrane(polyethylene sheet) **38.** gravel 25cm **39.** precast concrete threshold **40.** concrete slab with quartz powder **41.** phenolic fair-faced concrete slab **42.** plasterboard **43.** L 100.100.4 **44.** anodized aluminum door with thermal break and double glazing **45.** IPN 100

剖面详图 section detail

潘普洛纳某托儿所

Pereda Pérez Arquitectos

如今，对建筑结构的环保要求与对环境的关注紧密相关。

在潘普洛纳的这个托儿所项目中，其场地的价值是一个主要制约因素，没有这一点，这个项目不会很容易被理解。这个场地源于一个重新规划的项目，位于潘普洛纳郊区的边缘。这是一处特殊的三角形场地，其中作为分隔墙的两面侧墙是已建成的一排建筑物的后墙，第三面墙是长度最长的一侧，朝南，向入口处广场开放。

考虑到场地和项目所具有的极端条件，建筑师认为这个项目就是一座实验建筑，首先要根据建筑的功能布局加入设计的常识与逻辑，而且相较于其他极为复杂或繁多的方法，应优先利用太阳光。出于这个原因，建筑方案没有设计一座坐落在突出位置上的孤立建筑；建筑师将该项目理解为原有建筑的基座。

该项目从场地和项目本身出发，期望可以从对比的两种情况开始施工：第一种是建筑外部，外观以一种有序的方式被建筑师处理成一个抽象的基座，是模块化并经过分类的；另一种是建筑内部，内部空间被他们处理成不均匀的、有渗透性的、友好且明亮的。

该项目基本上被设计成五个教育区域和一个大的公共空间，公共空间将之前提到的空间组织起来，并且可以在天气条件不适合使用两个室外运动场时，用作娱乐空间，公共空间还要求有拱廊。另外，项目中还设置了容纳行政办公室、设备和设施的额外室内空间。

该项目位于场地的内部。两面支撑墙中间的空间被规划为服务空间，服务于公共区域、行政办公室和其他设施，它们围绕一个庭院设置，而庭院为建筑提供了必需的空间、阳光和通风。在分隔两面支撑墙的空间里，有建筑物的中央空间、过渡空间、玩耍空间、多功能空间，同时多功能空间直接与拱廊和运动场相连。

天窗是用来捕捉直射阳光的构件，它们被重叠设置，并且根据其在房间进深的布局来增加自身的高度，目的是使光线能照射到室内，而不是改善分隔墙的视觉冲击效果，因为分隔墙与体量非常协调。庭院将非直射的光线分散开，改善照明状况，并在每个教育区域内营造出亲切的气氛。

建筑物有两个入口，主要入口几乎设在立面中央，是通过格状立面从广场进入建筑的唯一入口。这个入口通往拱廊和主要的室外运动场。

项目自设计开始，最终又回到了原点，其构思充分考虑到了环境问题。毫无疑问，这种设计一方面在很大程度上使用地热资源作为可再生的清洁能源，另一方面，将上层建成花园，除了提高建筑物的热工性能，也因其是一群较高建筑物之间的单层建筑而提高了视觉效果。

Nursery School in Pamplona

The environmental commitment required nowadays to the architecture is strongly linked with the contextual concern.
In the nursery school project in Pamplona the value of the site is a main constraint, without it this project would not be easy to understand. The plot, result of a reallocation project, is located on the edge of a suburb in Pamplona. The geometry of such plot had practically a triangular shape, where two of its sides had the nature of dividing walls corresponding to the rear of a row of buildings, while the third side, and with the longest length, was orientated to the south, open to the square where the access is built.
Due to the extreme conditions of the plot and the program the architects thought that it was an exercise where firstly the common sense and the logic should be imposed in relation with the lay-out of the uses of the building and the priority of exploiting the sunlight prior than any other approach of highest complexity or exuberance. For that reason, the proposal was not adjusted to an isolated construction in an aggressive location; the architects understood it as the plinth of the preexisting buildings.
The project from the site and the program expected to work from the contrast adapting to the two situations; on one side the outside is treated as an abstract plinth, in an ordered manner, modular and categorical. And on the other side, the inside of the section

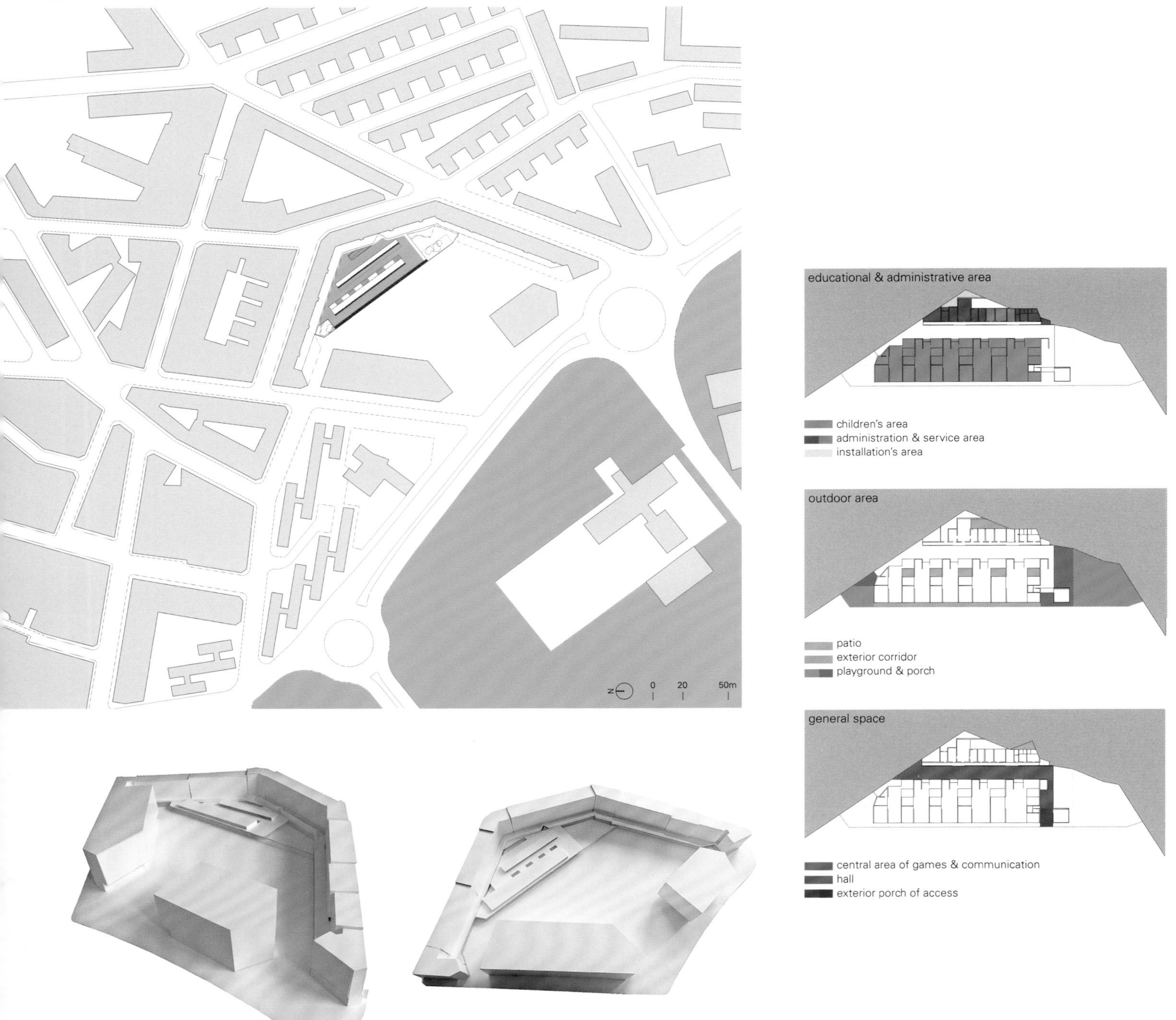

treated as uneven, permeable, friendly and luminous.
The program was basically adjusted to five educational modules, a big general space to organize the prior mentioned spaces and to allow to be used as entertainment space when the climatologic conditions prevent to enjoy the two outdoors playgrounds, with its arcades also requested, besides additional indoor spaces for administration, services and facilities.
At the internal part of the plot it is placed and spaces between two supporting walls are planned as the spaces of service to public, administration and facilities organized around a courtyard which supplies them with primary necessities, sunlight and ventilation. Between both spaces separating two supporting walls, it is placed the central space of the building, space of transits, space to play, multi-uses space which at the same time is communicated directly with the arcades and playgrounds.
The skylights, as mechanisms to capture the direct sunlight, are overlapped and increased in height according to its arrangement in the depth of the room with the mission of making the light penetrating at the same time than improving the visual impact of the dividing wall as it remains harmonized by those volumes. And the courtyards, diffuser of indirect light, reinforce the lighting and generate an intimate atmosphere in each of the educational modules.
The building shows two accesses, the main one, placed almost in the centre of the facade, is the only point in the lattice which allows the access to the building from the square. This entry allows access to the arcade and the main outdoor playgrounds.
Finally the project since its design, and coming back to the beginning, was conceived under a strong environmental concern. And without any doubt, such orientation collaborates remarkably on one side using the geothermal resource, as renewable and clean energy. And on the other, making the upper deck as a garden area which besides improving the thermal behavior of the building improves the visual impact since it is a single floor construction among higher buildings. Pereda Pérez Arquitectos

项目名称：Nursery School, Pamplona
地点：Alfredo Floristán Square, Spain
建筑师：Pereda Pérez Arquitectos
合作者：Teresa Gridilla Saavedra
施工方：Construcciones Guillén O.P.
甲方：Council of Pamplona
用地面积：2,457m²
建筑面积：1,630m²
总楼面面积：1,630m²
设计时间：2009.9
施工时间：2011.6
摄影师：©Pedro Pegenaute

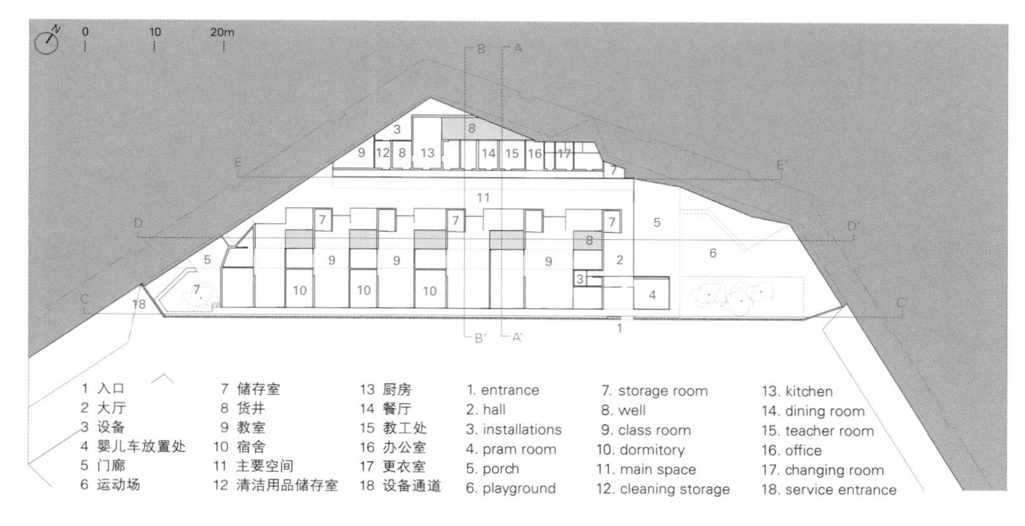

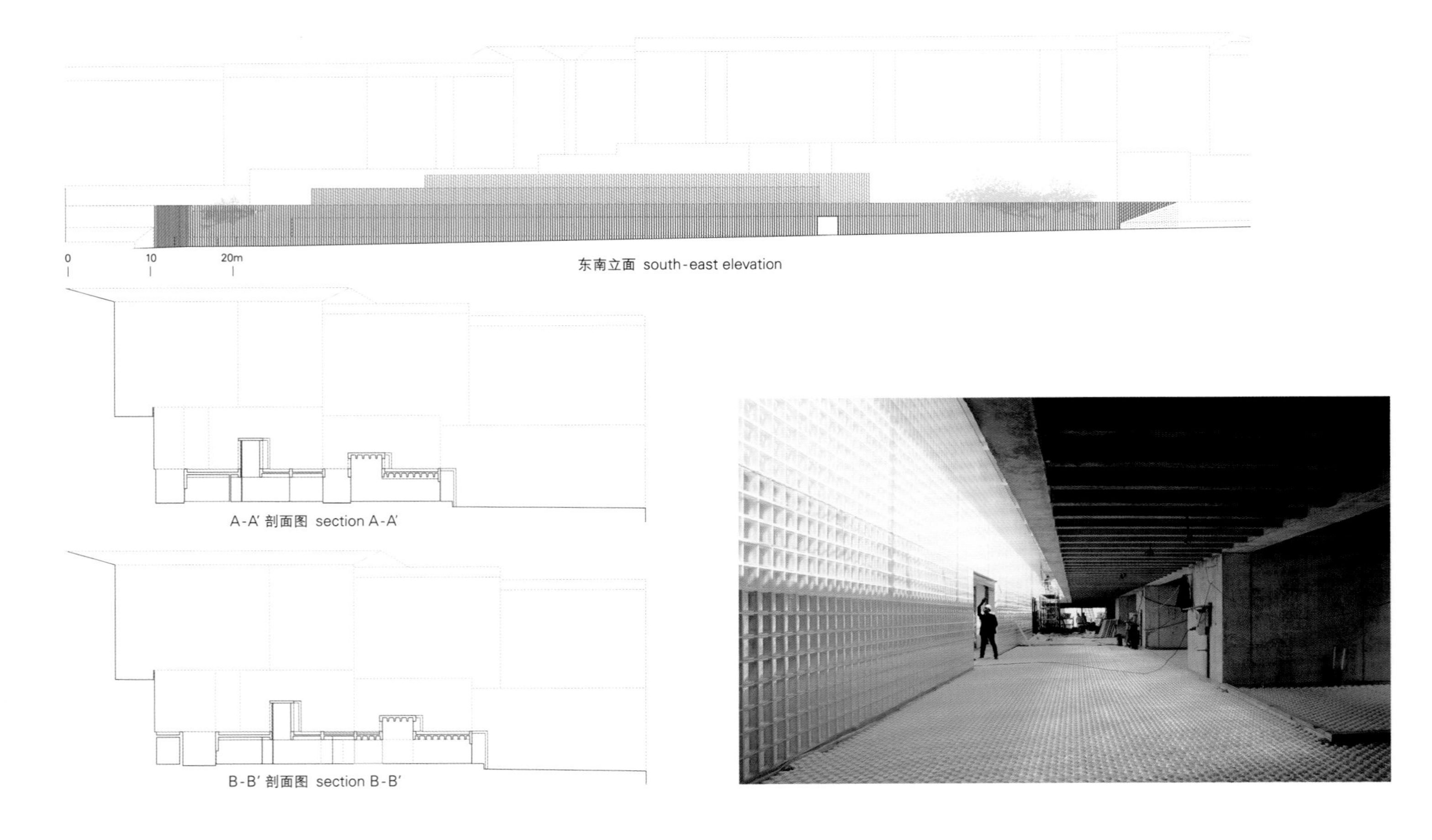

东南立面 south-east elevation

A-A′ 剖面图 section A-A′

B-B′ 剖面图 section B-B′

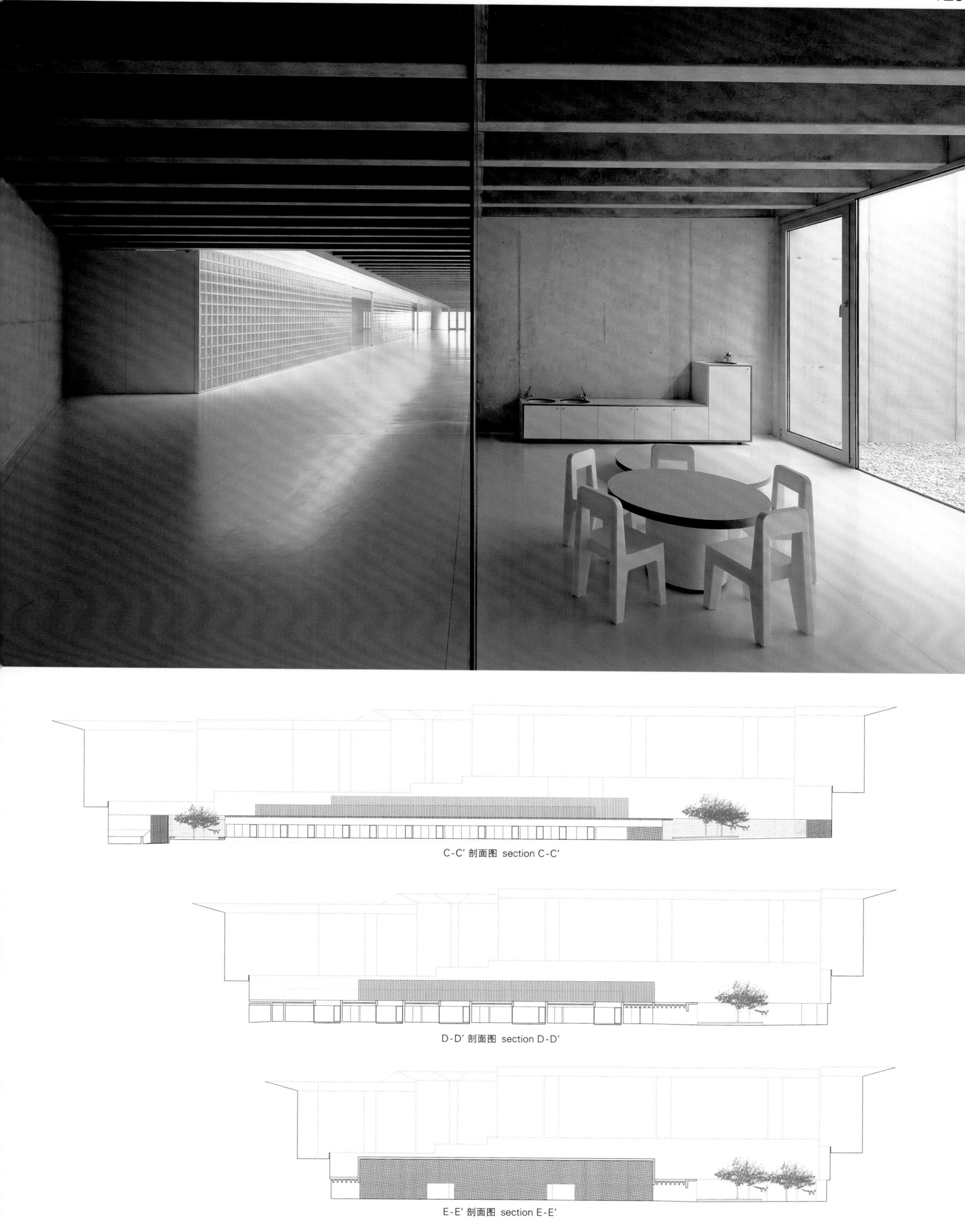

C-C′ 剖面图 section C-C′

D-D′ 剖面图 section D-D′

E-E′ 剖面图 section E-E′

1. concrete wall
2. concrete foundation
3. waterproof paint
4. waterproof sheet "delta drain"
5. sand crushed
6. PVC sheet
7. PVC pavement 2mm
8. leveling mortar 20mm
9. mortar bed
10. radiating floor
11. concrete bed
12. aluminium subframe 10.30.2
13. aluminium frame
14. galvanized steel 1.5mm
15. polished concrete floor
16. reinforced concrete cube
17. waterproofing EPDM
18. insulation extruded polystyrene 6cm
19. retaining and drainage
20. PVC cove base moulding
21. protection mortar
22. galvanized steel edging 1.5mm
23. plant substrate sedum
24. prefabricated concrete slab
25. compression layer
26. threaded joint
27. "I" prefabricated concrete
28. rock wool 50kg/m³
29. insulation 10mm
30. plasterboard cladding 15mm
31. lineal diffuser Trox AF
32. lightweight concrete
33. foscurit screen
34. expanded polystyrene joint
35. waterproof mdf 16mm
36. connector ø16mm
37. vapour barrier
38. reinforced concrete beam
39. laminated glass 6+6
40. "U" galvanized steel 150.20.1,5
41. aluminium angular 60.50.1,5
42. "L" aluminium 30.15.1
43. PVC lining e: 2mm
44. delta drain waterproofing
45. extruded polystyrene insulation 8cm
46. "U" aluminium 60/30/1.5
47. cristher discovery lamp
48. aluminium profile 15.50.1
49. PVC ø50mm
50. reinforced concrete beam
51. plasterboard partition 11.5cm
52. pine batten 50.25mm
53. plasterboard ceiling e: 1.2cm
54. celenit a panel 35mm
55. steel PL 68|218 8mm
56. reinforced concrete lintel
57. steel 10mm
58. steel profile 100.80.8
59. steel profile 160.80.8
60. sprayed intumescent mortar
61. glass brick vetroes C_smooth
62. insulation mortar 20mm
63. prefabricated concrete slab 20+5
64. pottery cinca 10.10
65. galvanized steel 80.10
66. galvanized steel profile 60.30
67. reinforced concrete wall
68. existing party wall
69. reinforced concrete slab

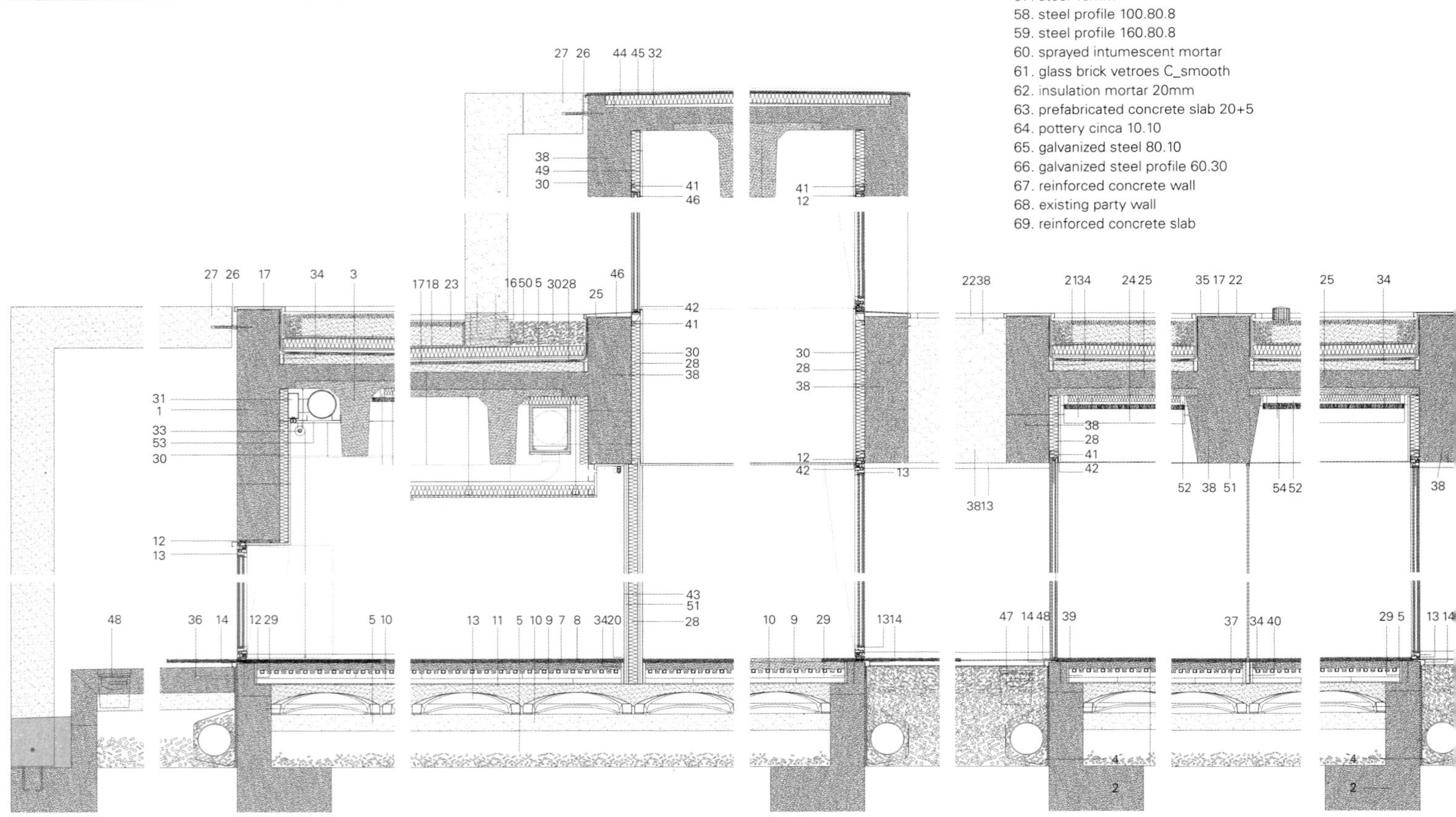

B - B′ 剖面详图 detail section B - B′

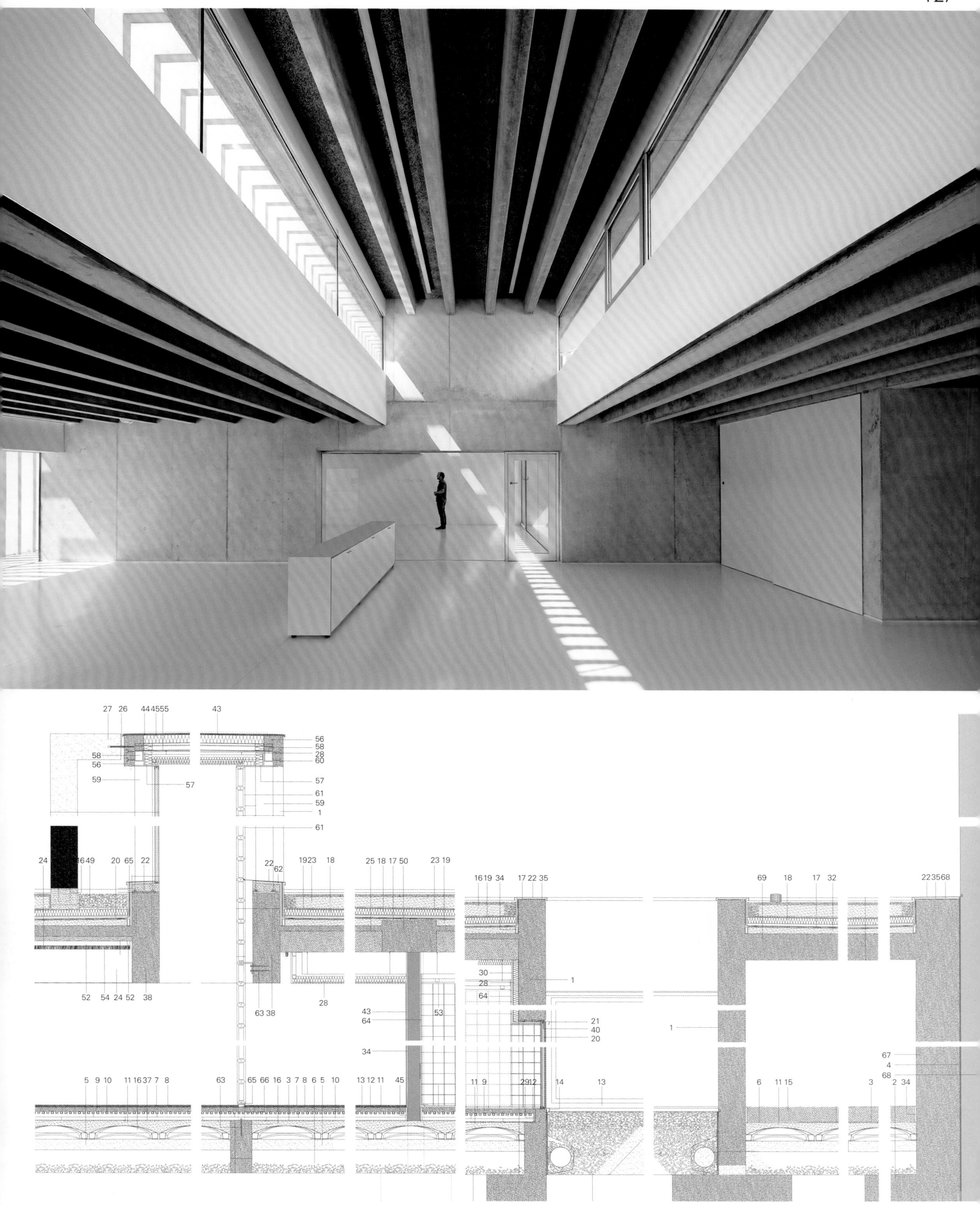
27 26 44 45 55
43
56
58
28
60
58
56
59
57
57
61
59
1
61
24
16 49
20 65 22
22
62
19 23 18
25 18 17 50
23 19
16 19 34
17 22 35
69 18
17 32
22 35 68
52 54 24 52 38
63 38
28
30
28
64
1
43
64
53
21
40
20
34
1
67
4
68
5 9 10
11 16 37 7 8
63
65 66 16 3 7 8 6 5 10
13 12 11
45
11 9
29 12
14
13
6
11 15
3
2 34

Mário Sequeira美术馆

Atelier Carvalho Araújo

当在一个传承了周围环境特色的地方嵌入一座建筑时，建筑师应通过保留的方式来尊重该场地，这不仅仅指代保留该地的景观，保留该地区所固有的习俗和姿态也是至关重要的。

为了尊重原有建筑并使自然占主导地位，建筑师将新美术馆建筑恰当地设置在斜坡上，该建筑的独特用途和功能空间被设置在地下。建筑外立面所流露出的悄无声息的视觉效果以及规模的二元性更加激发了参观者对于室内的好奇心——对室内的好奇程度远大于对建筑外表的好奇。

参观者越深入到建筑内部，对空间规模以及灯光和色彩的对比度的感觉就越强烈。项目、入口区域以及交通流线的简洁性都是为了凸显出周围的自然景观。位于低处的室内空间向东面开放，并与室外空间建立了一种视觉联系。

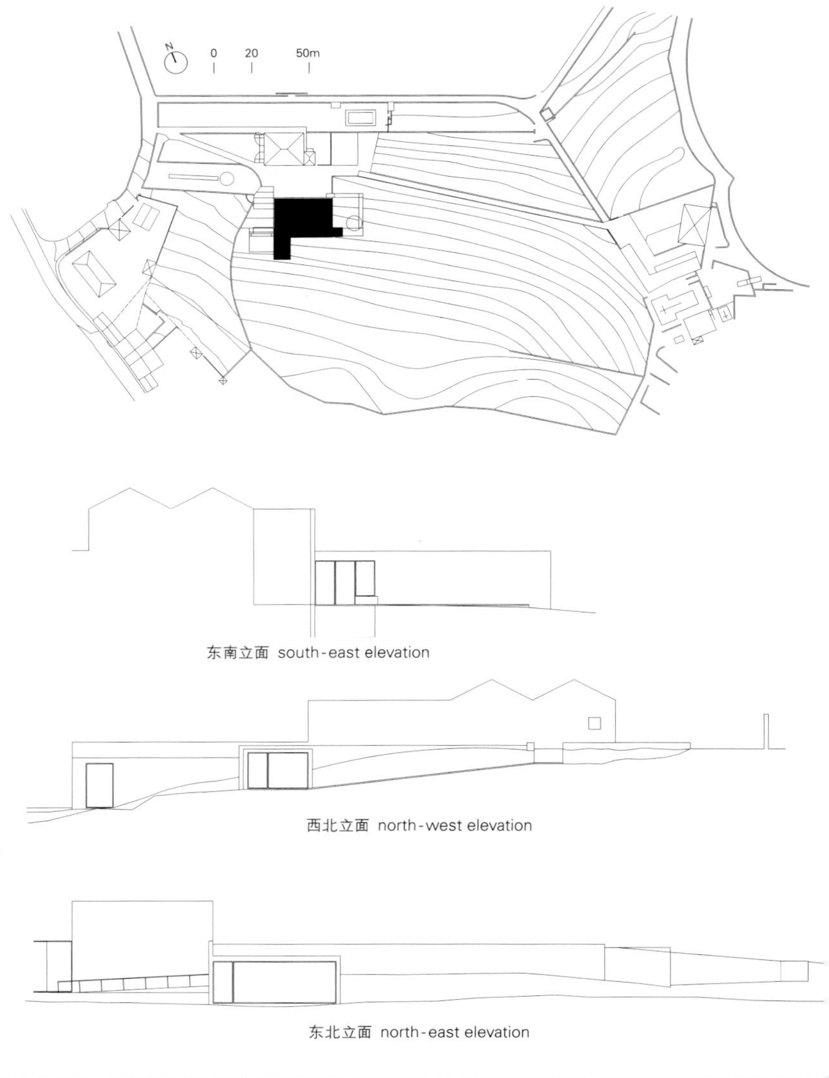

东南立面 south-east elevation

西北立面 north-west elevation

东北立面 north-east elevation

Mário Sequeira Gallery

When inserting the building in an area where cultivated identity of the surrounding environment inherited, architects should respect the site with a strong sense of preservation, not in the sense of the landscape, but for the habits and attitudes inherent in the "feel" of the place is critical.

To remain subordinate to the existing and to yield to the prevailing nature, the new art gallery building was laid modestly on the slope with its specific uses and functions inserted under the ground. The visual silence and duality of scale maintained on the exterior double up the curiosity of the visitors about the interior – less on the surface and more inside.

As the visitor approaches deeper into the building, the feeling of the space scale along with the contrasts of light and color grows. Simplicity of the program, zones of access and circulation, was designated to yield to the surrounding nature. The interior space laid low is open to the east side, establishing a visual relation with the outer space. Atelier Carvalho Araúj

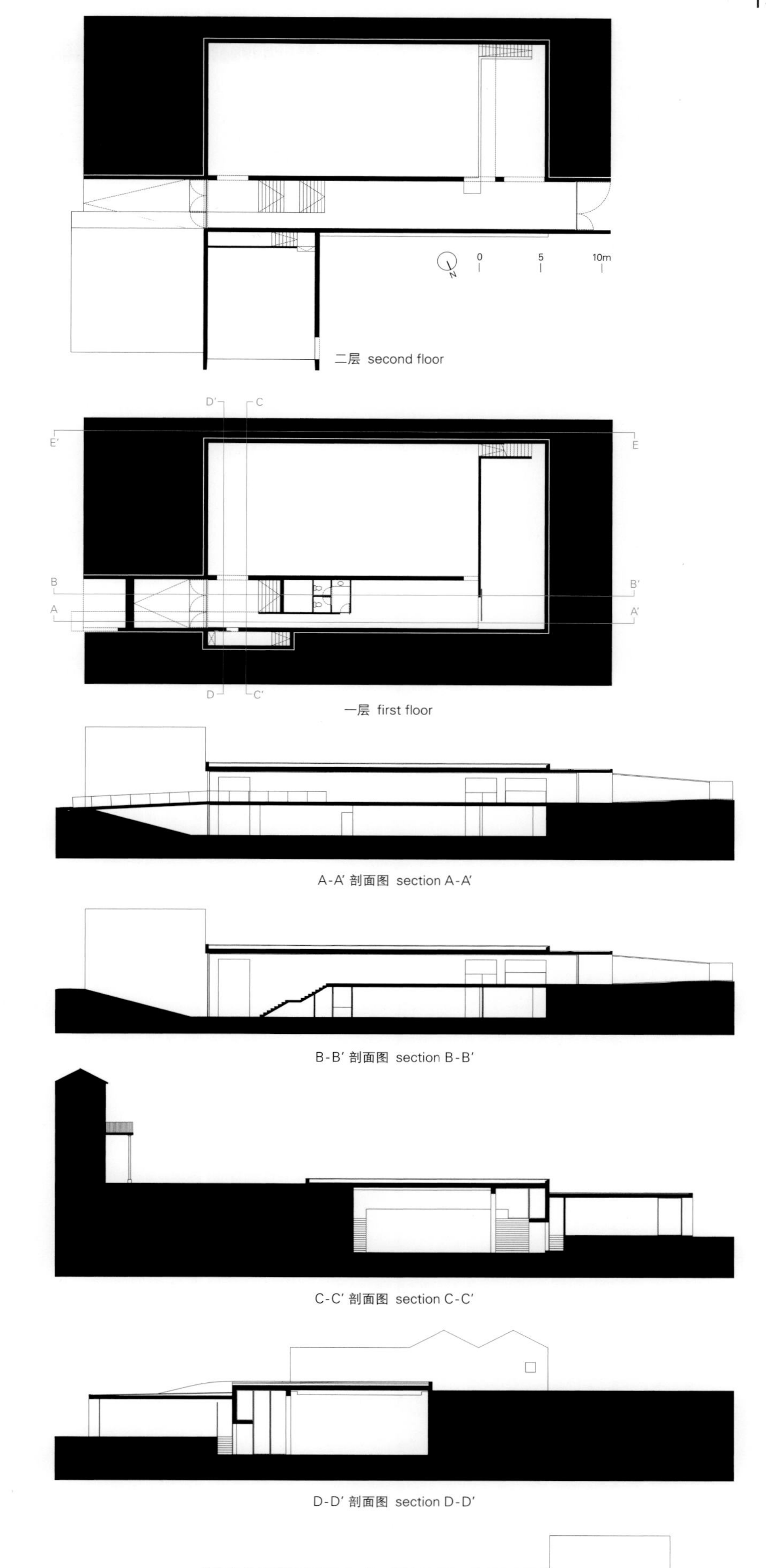

项目名称：Galeria Mário Sequeira
地点：Braga, Portugal
建筑师：J M Carvalho Araújo Arquitectura e Design, S.A.
项目团队：José Manuel Carvalho Araújo, Nuno Capa, Raul Carvalhais
用地面积：35,870m²
建筑面积：788m²
总楼面面积：845m²
摄影师：©Pedro Lobo (courtesy of the architect)

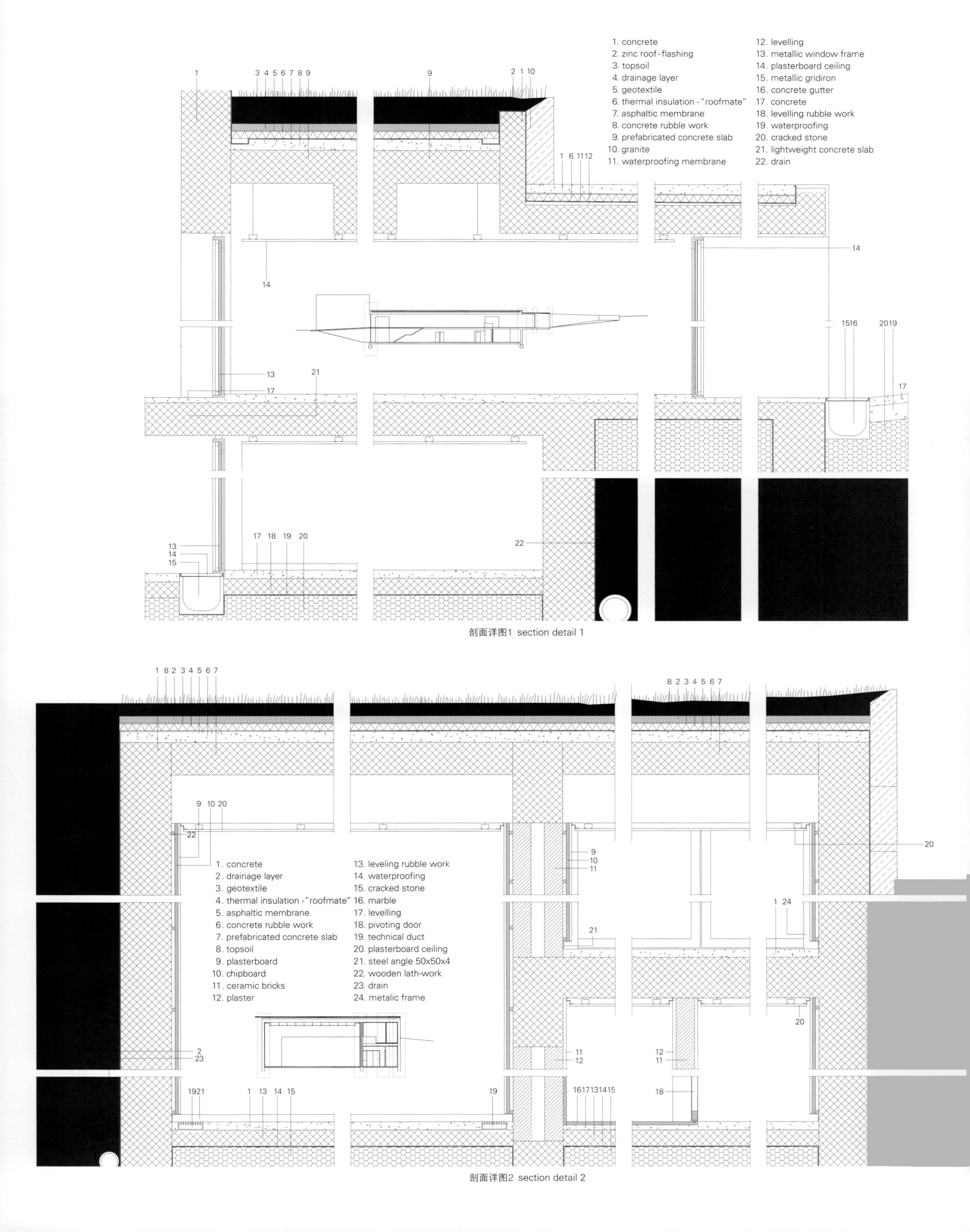

剖面详图1 section detail 1

剖面详图2 section detail 2

意大利—委内瑞拉中心的新服务大楼

Roberto Puchetti

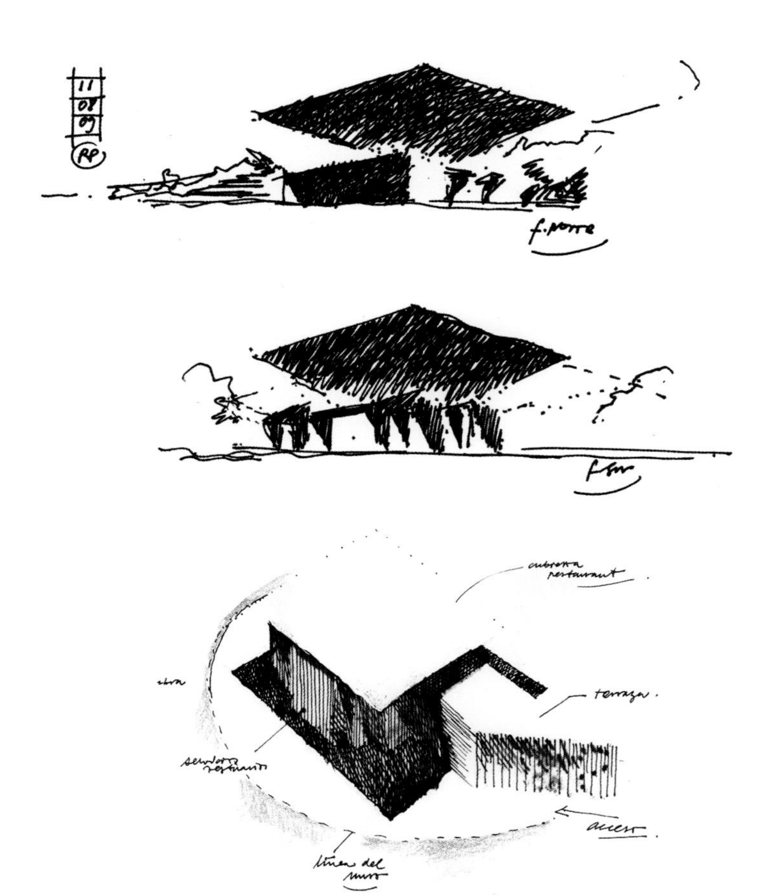

意大利—委内瑞拉中心的新服务大楼位于加拉加斯东南的小山上，与该中心的基础建筑毗邻，这些基础建筑由获得世人盛誉的建筑师Antonio Pinzani在20世纪60年代和80年代设计，具有独特的几何形状特征：薄薄的混凝土双曲面（HYPAR）构成该中心主要空间的屋顶，清晰地表达出Pinzani想要打造适合热带气候的空间的意图。这次的方案试图尊重他原来的设计。设计过程一直被周围郁郁葱葱的热带植物形成的阴影这一“先决条件”影响。当委内瑞拉的国际知名画家Armando Reverón试图在他的画中强调加勒比地区的耀眼光芒时，他从一块白色画布开始着手，“先决条件”就是先画阴影。那么是否可以构想这样一座建筑：利用照射在一个暗面上的光线来突出阴影？

设计过程基于将光线—阴影的关系转变阴影—光线，也就是说，由绘制“先决条件”所需的阴影开始，然后绘制对应这些阴影的体量和表面。这样设计的结果是形成了一个屋顶，不但成为各部分的主导，还将它们组织到一起，同时强调了具有不同功能的体量之间的空间流动性。屋顶同时成为遮阳装置以及美学元素，因其拱腹带有多个不同的面，而形成不同的阴影强度。

在整个设计过程中，建筑师从环境影响和经济两方面特别注意了项目的可持续性。当支柱的数量从12根减少到8根时，整个结构变得非常高效，成本在原预算的基础上降低了30%。另外由于场地南侧采用了弧形挡土墙，减少了混凝土和钢筋的使用量，又节省了25%的成本。在能源消耗方面，由屋顶带来的阴影有可能节省多达每小时22.5kW的用电量，有助于避免空调系统的使用，并将室内温度降低6°C。六月份（最热的月份）建筑物最大限度地被保护起来，可以免受太阳辐射的损伤，而在十二月，阴影覆盖了95%的空间。

屋顶由四个酒杯状结构组成，每一个结构都有一根柱子支撑，作为收集雨水的漏斗。这个简单的结构每月都可以收集到数量可观的雨水，从而节省了数百升用水。收集到的水用于冲厕所、清洗地板、灌溉植物和清洁活动区域。预算不足意味着无法使用技术先进的环境控制系统。建筑师仅仅依靠建筑自身及其构件（建筑朝向、自然通风、建筑构件的尺寸和比例）来适应天气条件。

Italian-Venezuelan Centre New Services Building

The New Services Building for the Italian-Venezuelan Centre, located on the south-east hills of Caracas, sits next to the centre's foundational buildings, designed by the never praised enough architect Antonio Pinzani during 60's and 80's. A particular geometry characterizes those early buildings: thin double curved concrete surfaces (HYPAR) roof the main spaces of the centre, making clear Pinzani's intention to create spaces sensible to the tropical weather. The proposal attempts to pay homage to his vision.

The design process was stimulated by the "a priori" value given in the tropics to the shadows. When internationally renowned Venezuelan painter Armando Reverón attempted to highlight the blinding light of the Caribbean in his paintings, he started working them out from a white canvas, "a priori" stained with shadows. Is it then possible to think on an architectural composition that attempts to highlight the shadows by working out with lights staining a dark plane?

The design process is based on the inversion of the relationship Light-shadow into Shadow-light, that is, it began by drawing the shadows needed by a priori, followed by the volumes and surfaces that correspond to such shadows. The result is a roof which dominates and organizes the composition while emphasizing the

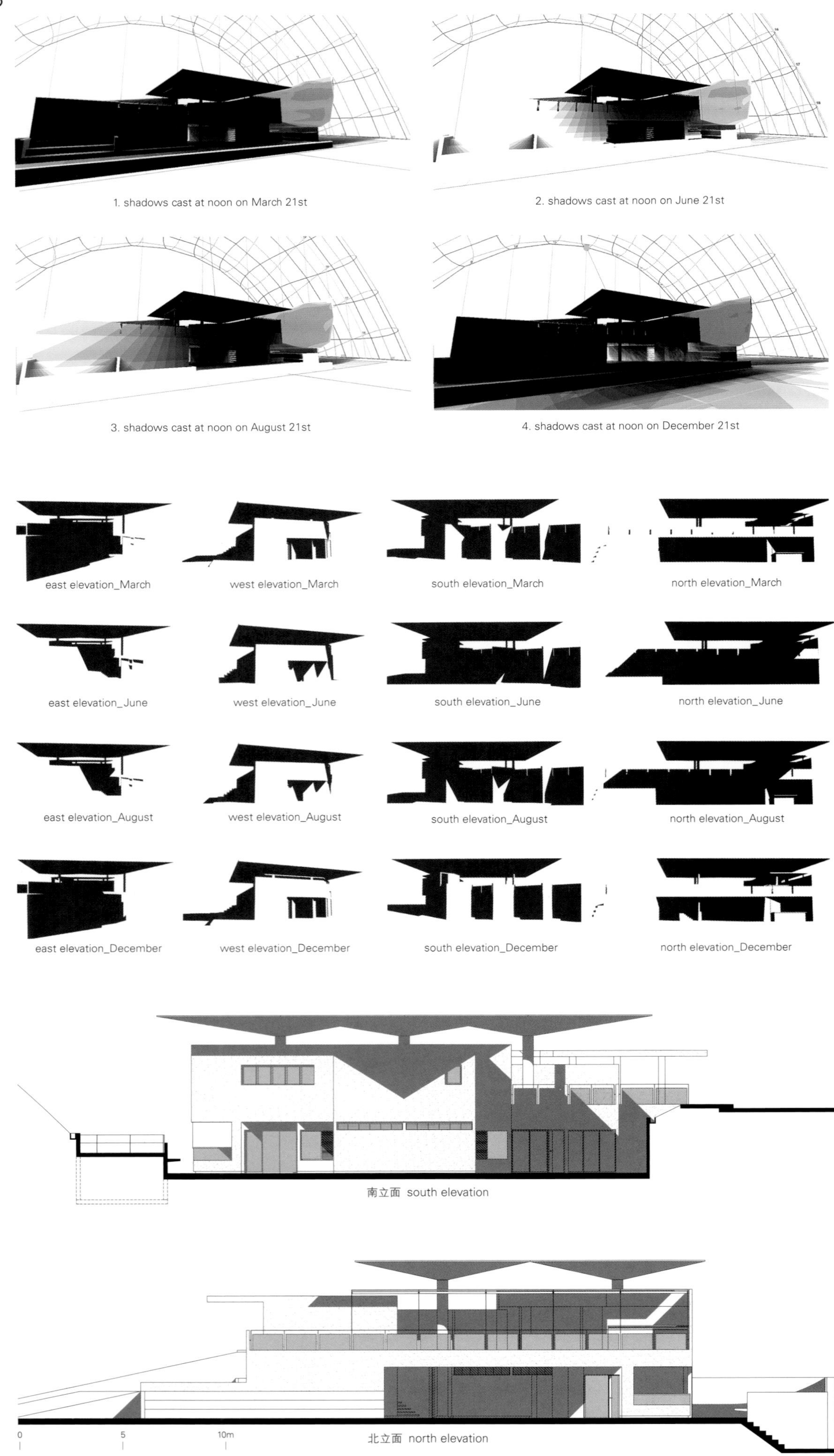

1. shadows cast at noon on March 21st
2. shadows cast at noon on June 21st
3. shadows cast at noon on August 21st
4. shadows cast at noon on December 21st
east elevation_March
west elevation_March
south elevation_March
north elevation_March
east elevation_June
west elevation_June
south elevation_June
north elevation_June
east elevation_August
west elevation_August
south elevation_August
north elevation_August
east elevation_December
west elevation_December
south elevation_December
north elevation_December
南立面 south elevation
0
5
10m
北立面 north elevation

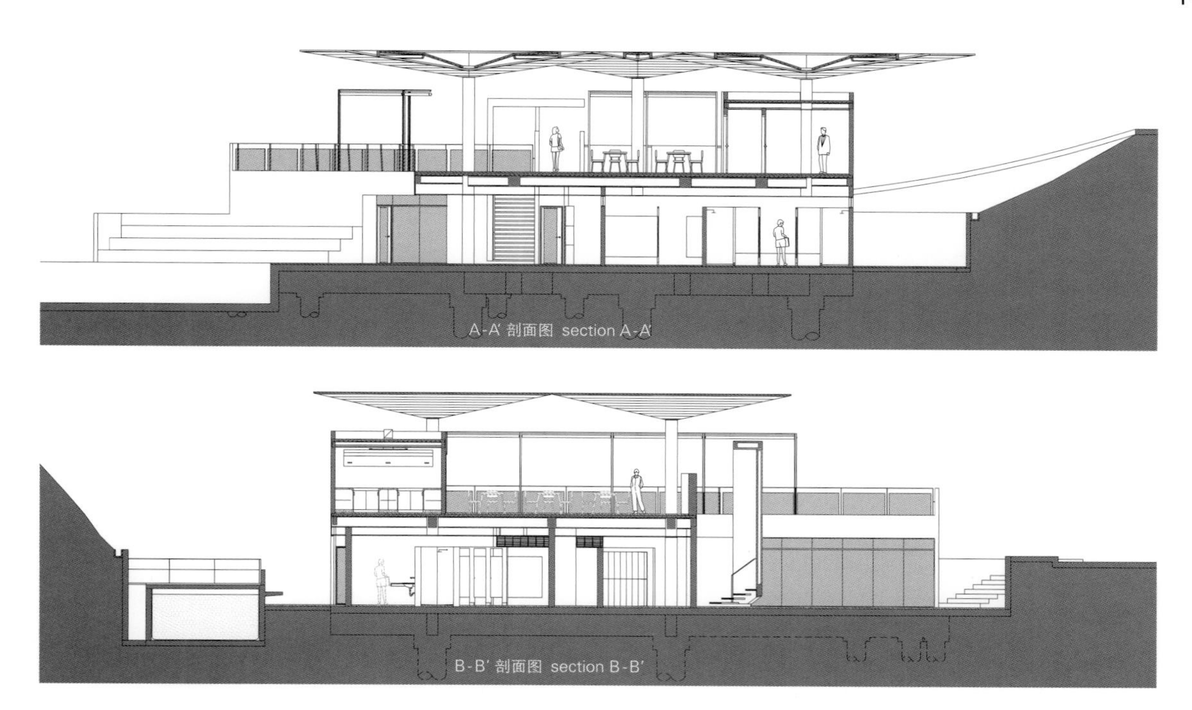
A-A' 剖面图 section A-A'
B-B' 剖面图 section B-B'

fluidity of the spaces between the volumes containing the program. The roof becomes both a shadowing device as well as an aesthetic element which creates different shadow intensities, due to the faceted geometry of its soffit.

During the entire design process special attention was given to the sustainability of the project, both in terms of the environmental impact and economy of means. The structure became extremely efficient when the number of pillars was reduced from 12 to 8, achieving a cost reduction of 30% on the original budget. Another 25% of savings was achieved on the amount of concrete and steel due to the arc-shaped retaining wall to the south of the site. In terms of energy consumption, a reduction up to 22.5kW/h was possible as result of the amount of shadows produced by the roof, which helps avoiding the use of air conditioning system and reduces the interior temperature by 6 degrees. In June (the hottest month) the building is protected enough from solar radiation and in December the shadow covers up to 95% of the space.

The roof is made out of four chalices, each one supported by a column which acts as a funnel collecting the rain water captured. This simple system helps save hundreds of liters by collecting a considerable amount of water per month. The collected water is used for WC cisterns, cleaning of floors and irrigation of green and sports areas. The scarce budget meant that the use of technologically sophisticated systems of environmental control was out of reach. The architect relied solely on architecture and its elements (solar orientation, natural ventilation, dimensions and proportions of architectural elements) to negotiate with weather constraints.

Yacira Blanco

项目名称：Italian-Venezuelan Centre New Services Building
地点：Caracas, Venezuela
建筑师：Roberto Puchetti
项目团队：Yesenia Di Sabatino, Fernando Blanco, Andrés Cova, Karina Rodriguez
结构工程师：Manuel Ramírez
供水和排水工程：Manuel Ramírez
机械和天然气工程师：Ricardo Ortega
电气工程师：Roberto De Adessis
甲方：Centro Italiano Venezolano, A.C.
施工公司：Constructora Proyectos La Torre, C.A.
建筑面积：465m²
总楼面面积：465m²
材料：steel, render, wood, glass
设计时间：2009 施工时间：2010 — 2011

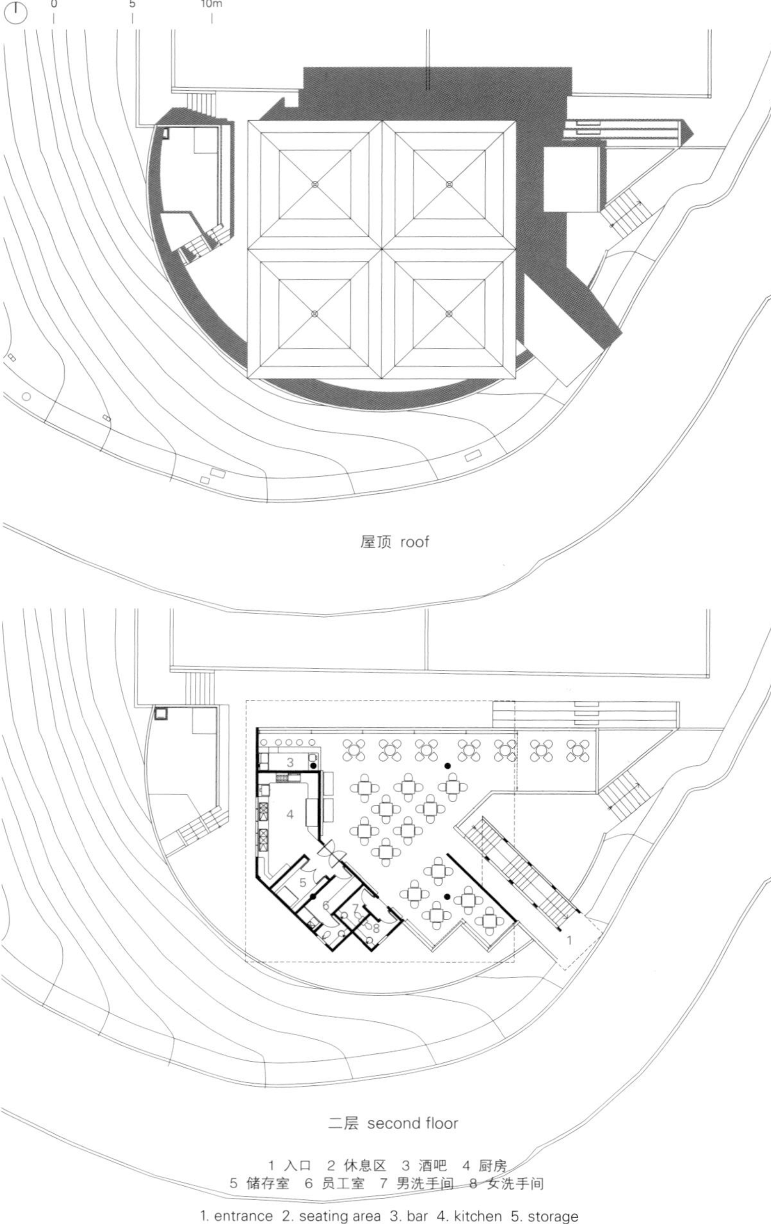

屋顶 roof

二层 second floor

1 入口 2 休息区 3 酒吧 4 厨房
5 储存室 6 员工室 7 男洗手间 8 女洗手间

1. entrance 2. seating area 3. bar 4. kitchen 5. storage
6. employees room 7. gents bathroom 8. ladies bathroom

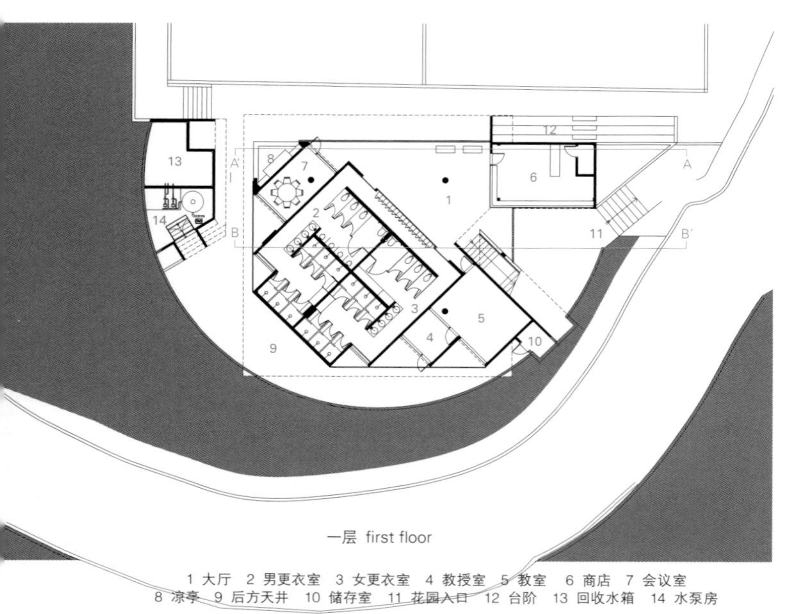

一层 first floor

1 大厅 2 男更衣室 3 女更衣室 4 教授室 5 教室 6 商店 7 会议室
8 凉亭 9 后方天井 10 储存室 11 花园入口 12 台阶 13 回收水箱 14 水泵房

1. lobby 2. gents changing room 3. ladies changing room
4. professor room 5. class room 6. shop 7. meeting room 8. kiosk 9. rear patio
10. storage 11. garden access 12. steps 13. recycled water tank 14. pump room

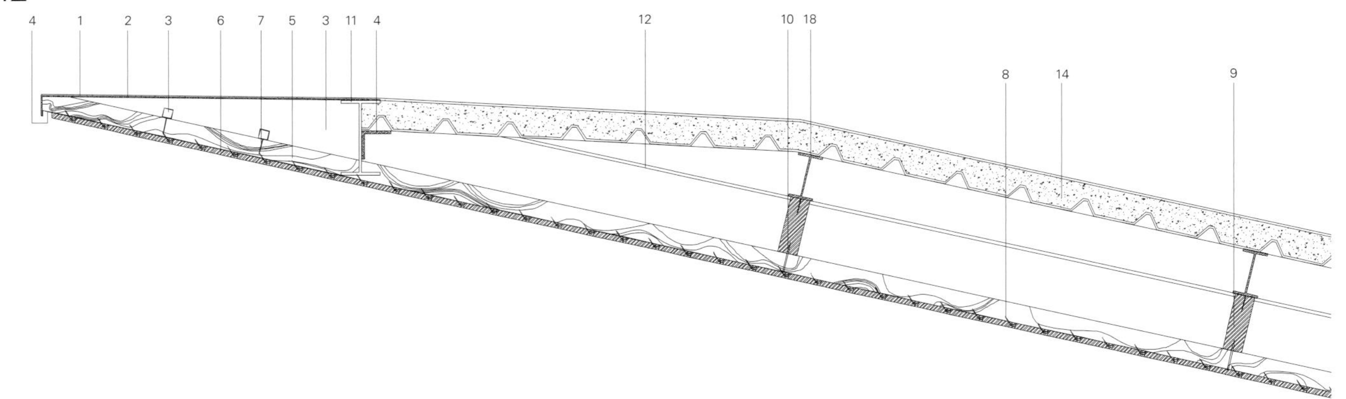
4
1
2
3
6
7
5
3
11
4
12
10
18
8
14
9

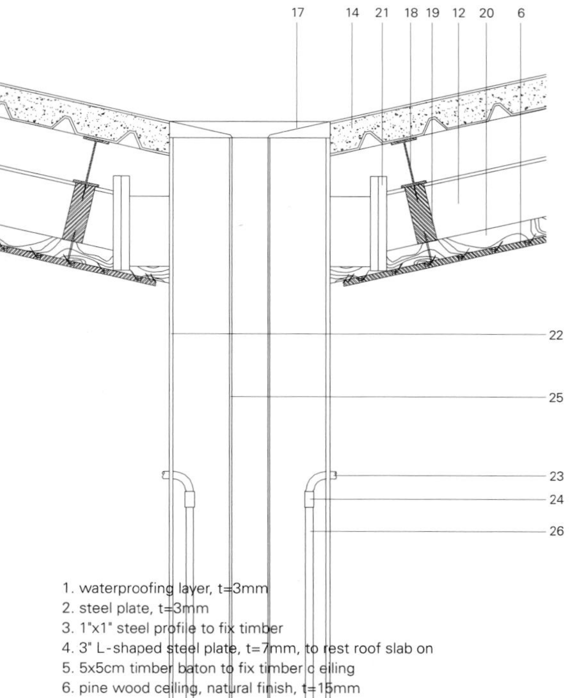

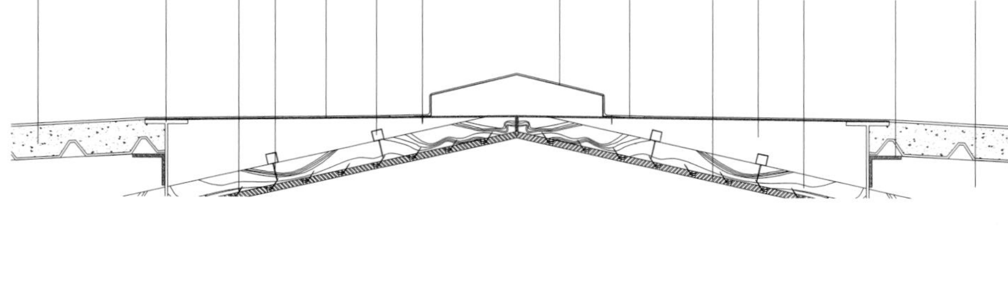

1. waterproofing layer, t=3mm
2. steel plate, t=3mm
3. 1"x1" steel profile to fix timber
4. 3" L-shaped steel plate, t=7mm, to rest roof slab on
5. 5x5cm timber baton to fix timber ceiling
6. pine wood ceiling, natural finish, t=15mm
7. nail welded to 1"x1" steel profile to anchor timber joist
8. nail to fix timber ceiling to joist
9. 5x15cm timber baton to fix joist to beam
10. nail welded to beam, for anchoring of 5x15cm timber baton
11. IPN 200 steel beam
12. IPN 200 slanted steel beam
13. 40cm round section steel column
14. concrete light mix on corrugated steel slab(Losacero), t=8cm
15. aluminium plate fixed over roof expansion joints
16. anchor fixing between aluminium and mental plate
17. asphaltic layer for waterproofing, t=3mm
18. IPE 120 steel beam
19. 5x15cm timber baton for fixing of timber beam to joists
20. 5x5cm timber baton as joist in timber dovetailing
21. 24x30cm metal plate, t=2mm, welded to IPN 200 slanted steel beam
22. steel column, diameter=40cm, t=12mm
23. lighting installation point in columns, height=3m
24. connection piece for light steel piping(E.M.T) , diameter=3/4",
25. steel pipe diameter=4", for rain water drainage
26. light steel pipe(E.M.T), diameter=3/4", for electric cabling

responsible use of natural resources

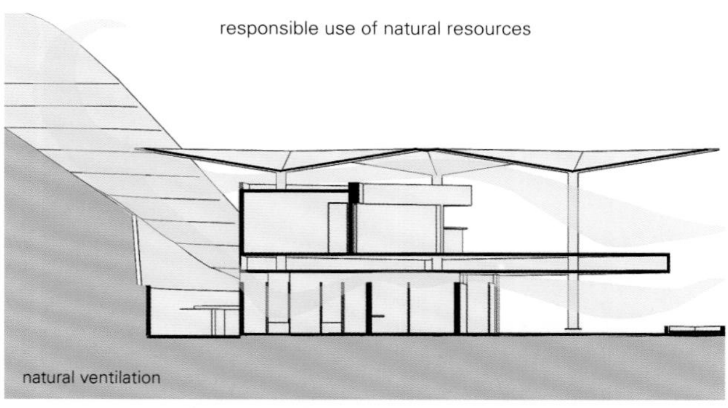

natural ventilation

- A study on wind's capacity to reduce the temperature inside the building was conducted using the software ecotech V5. Results proved wind can lower 5 to 6 degrees celsius on ground level and 3 to 4 degrees celsius on the second floor without the use of air-conditioning.
- In order to obtain this positive effect, the project must have a sequence of fluid and interconnected spaces where wind can move freely.
- Saving energy is significant(22.5kilowatts per hour), and maintenance costs of the equipments are also avoided.

wind recycling

- Rain falls abundantly in Caracas every year(1,279.5mm), and most of it is wasted running through the streets.
- During heavy rain, the funnel-shaped roof allows to collect 2800 liters per hour of water in the special section of the tank through pipes in pillars of the builing.
- This water will be used for cleaning and landscape watering.
- Recycling rain will save up to 56 US$ every month thanks to a lesser demand of water from the city's supplies.

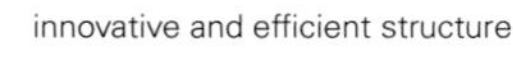

innovative and efficient structure

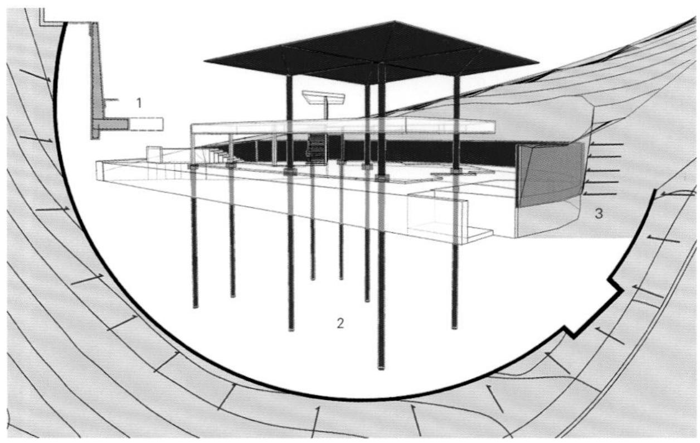

1. A retaining wall that describes an arch has a smaller concrete footing than a standard wall. This represents a reduction of more than 12m³ of concrete.
2. Eight 12m-long piles, reducing the usual quantity of material in half for this type of building, saves 38m³ of concrete.
Only four main pillars can support the weight from the second floor and main roof.
3. A curved retaining wall reacts as one structure to the pushing forces of the terrain. It requires less material than a standard wall to resist the same forces.

毛伊岛悬崖上的房子

Dekleva Gregoric Arhitekti

帆板船帆设计师Robert Stroj从欧洲搬到毛伊岛，来领导位于毛伊岛卡胡卢伊的Neil Pryde设计研究工作室。如果说毛伊岛南部海岸温和且适合欢度全包式假期，其北部海岸则有着强风和完美的海浪（最重要的因素），是勇敢的冲浪者的天堂。

它位于海洋中间，几乎远离所有东西。家在这样的环境中变得至关重要。除了仅仅作为一个家，这个房子同时也是业主的社交场所。晚上的活动就是所有人集中在一起烹饪，每位客人都认识到，烹饪不只是为了果腹，更是一种享受。因此，厨房和餐厅形成了房子的中心——一个位于封闭的私人空间中的流动空间。

这里有完美的海景、美丽的峭壁、强风和未受污染的自然景观。还有建造一座房子的空间吗？这对通常面对相当密集的城市环境的欧洲建筑师来说，是一个很不寻常的任务。建筑师在首次造访这里之前很难理解这里，但在那里待了几天之后，一切都变得容易理解起来。

设计理念在同一个屋顶下方形成了几所"房子"；屋顶作为一个折叠起来的木质甲板，在材料和结构上与周围景观融为一体。每所单独的"房子"都是一个U形体量，作为双人套间（卧室加浴室），并且面向完美的海景。

屋顶的设计概念与此处恶劣的气候（包括强烈的阳光和强劲的海风）密不可分。屋顶的面积是房子面积的两倍，因此被覆盖的室外空间和室内空间一样大。它不需要空调，因为其结构使得整个屋子可以进行穿堂式通风。折叠屋顶被小心地安装在这座U形体量的墙上，分割出各个独立的空间。

当地的材料被用于房子的饰面。所有的室内和室外的墙壁都用混合了沙滩上的沙子的石膏粉刷过，进一步强调出室内—室外融合的关系。地板和阳台、天花板甚至屋顶上使用了同样的重蚁木。另一方面，虽然这座房子是按照美国标准建造的，但是没有使用典型的混凝土砌体结构，以此反映出业主的欧洲血统。

建筑师愿意相信这座建筑物是在强大的施工过程监控下，在现场施工完成的。在施工过程中，他们不能够随心所欲地亲临现场，但是所有必要的监督和现场协调工作都由业主——一位工业设计师，非常仔细地完成。

Clifftop House Maui

Windsurf sail designer Robert Stroj moves from Europe to Maui to lead the design research studio of Neil Pryde in Kahului, Maui. If Maui's south coast is gentle and works for indulging all-inclusive holidays, its north coast is a rough surfer's paradise with strong winds and most important perfect waves.

It is in the middle of the ocean, far-off from almost everything. The home in such an environment becomes crucially important. Besides being just a home, this house works also as a social venue for the owners. The evening events are culinary blasts, where every guest realizes that cooking is not just necessity but more an obsession. Therefore the kitchen and the dining form the centre of the house – a fluid space between enclosed private volumes.

There are perfect ocean view, beautiful cliffs, strong winds and unspoiled landscape. Is there any space for a house? It was a very unusual task for European architects usually dealing with quite dense urban environment. It was hard to understand before the first visit, and easy to respond after some days spent there.

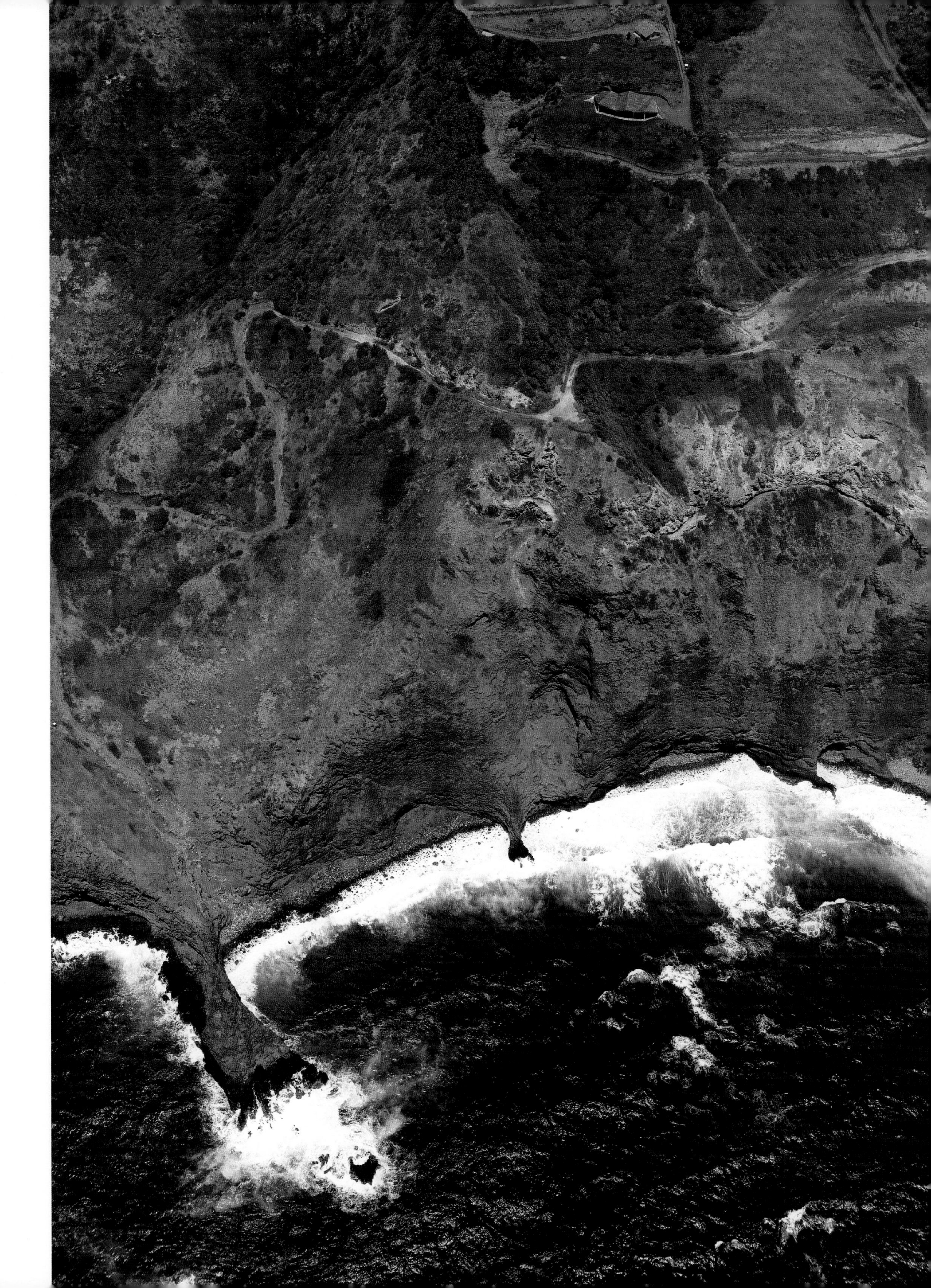

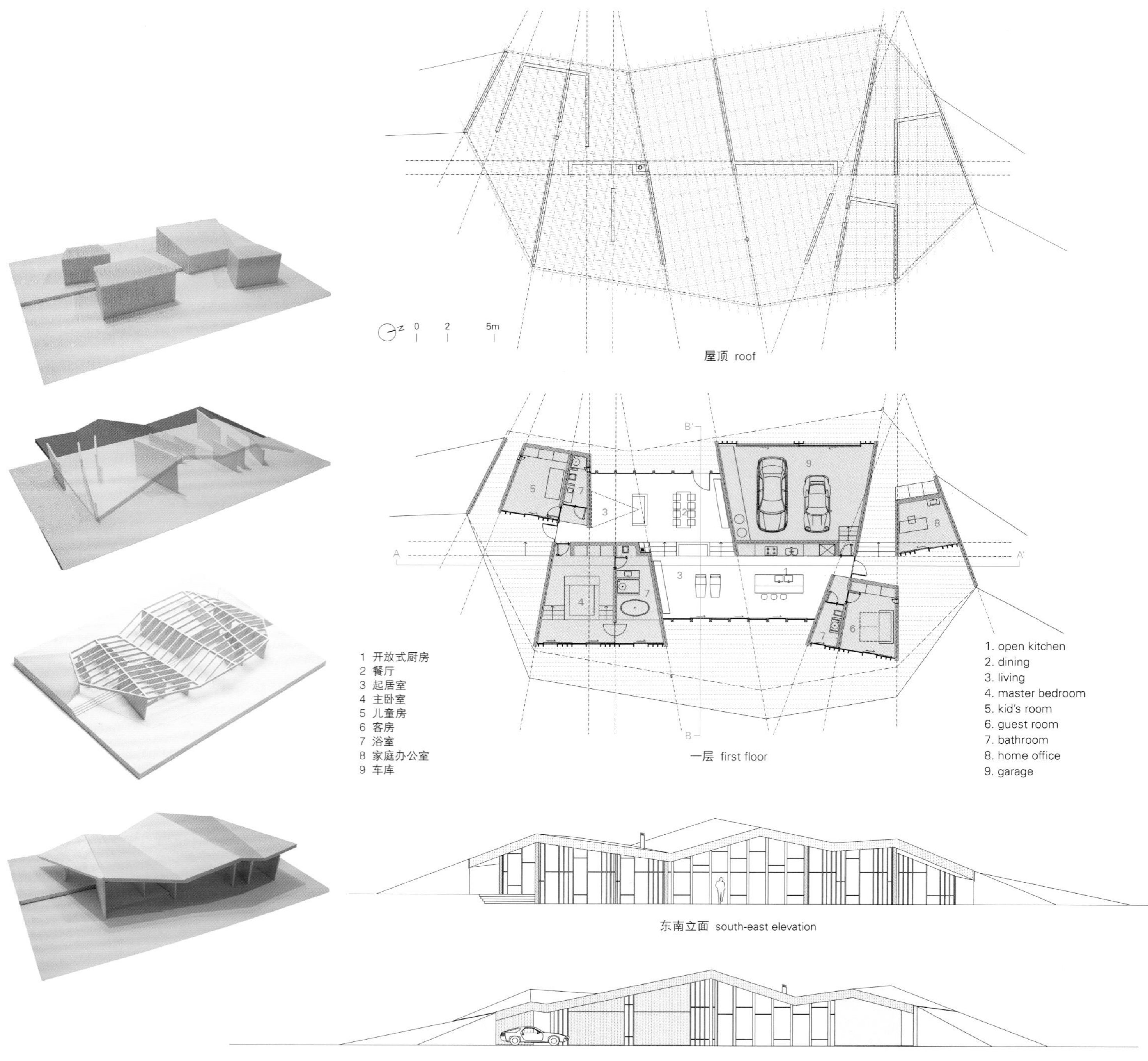

0 2 5m
屋顶 roof
1 开放式厨房
2 餐厅
3 起居室
4 主卧室
5 儿童房
6 客房
7 浴室
8 家庭办公室
9 车库
一层 first floor
1. open kitchen
2. dining
3. living
4. master bedroom
5. kid's room
6. guest room
7. bathroom
8. home office
9. garage
东南立面 south-east elevation
西北立面 north-west elevation

The concept defines several "houses" under a common roof which also serves as a folded wooden deck and materially and topologically integrates the house with the landscape. Each separate "house" is a U-shaped volume that works as a double room (bedroom plus bathroom), which opens up to the perfect ocean view. The roof concept is strongly related to the rough climate with loads of sun and strong ocean winds. The area of the roof is twice the size of the house, so the size of the covered outdoor space equals the size of the indoor space. It needs no air-conditioning, since it is cross-ventilated throughout. The folded roof is carefully attached to the walls of U-shaped volumes and defines specific spaces.

The local materials are used for the finishing of the house. The walls are rendered with the same plaster with beach sand indoor and outdoor and furthermore emphasize the smooth indoor-outdoor relationship. The same IPE wood is used for the floor plus terrace, ceiling and even the roof. On the other hand the house is for US standards atypically constructed out of concrete blocks, which just reflects the European origin of the owner.

The architects like to believe the architecture is done on-site with a strong control of the construction process. Here they were not able to visit the site while in process as often as the architects would want to, but all the essential supervision and site-coordination work was extremely carefully done by the owner – an industrial designer. Dekleva Gregoric Arhitekti

courtesy of the architect

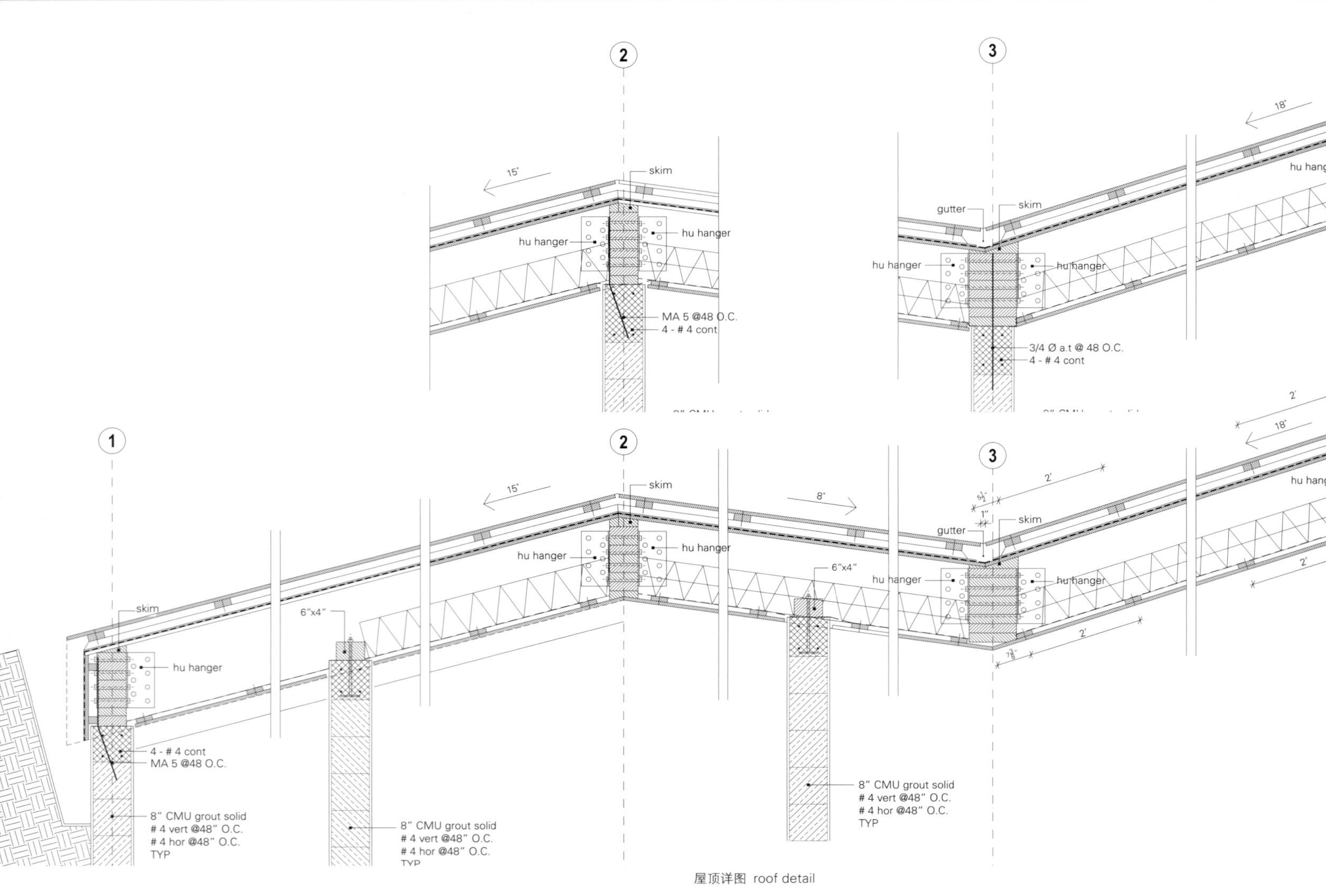

屋顶详图 roof detail

courtesy of the architect

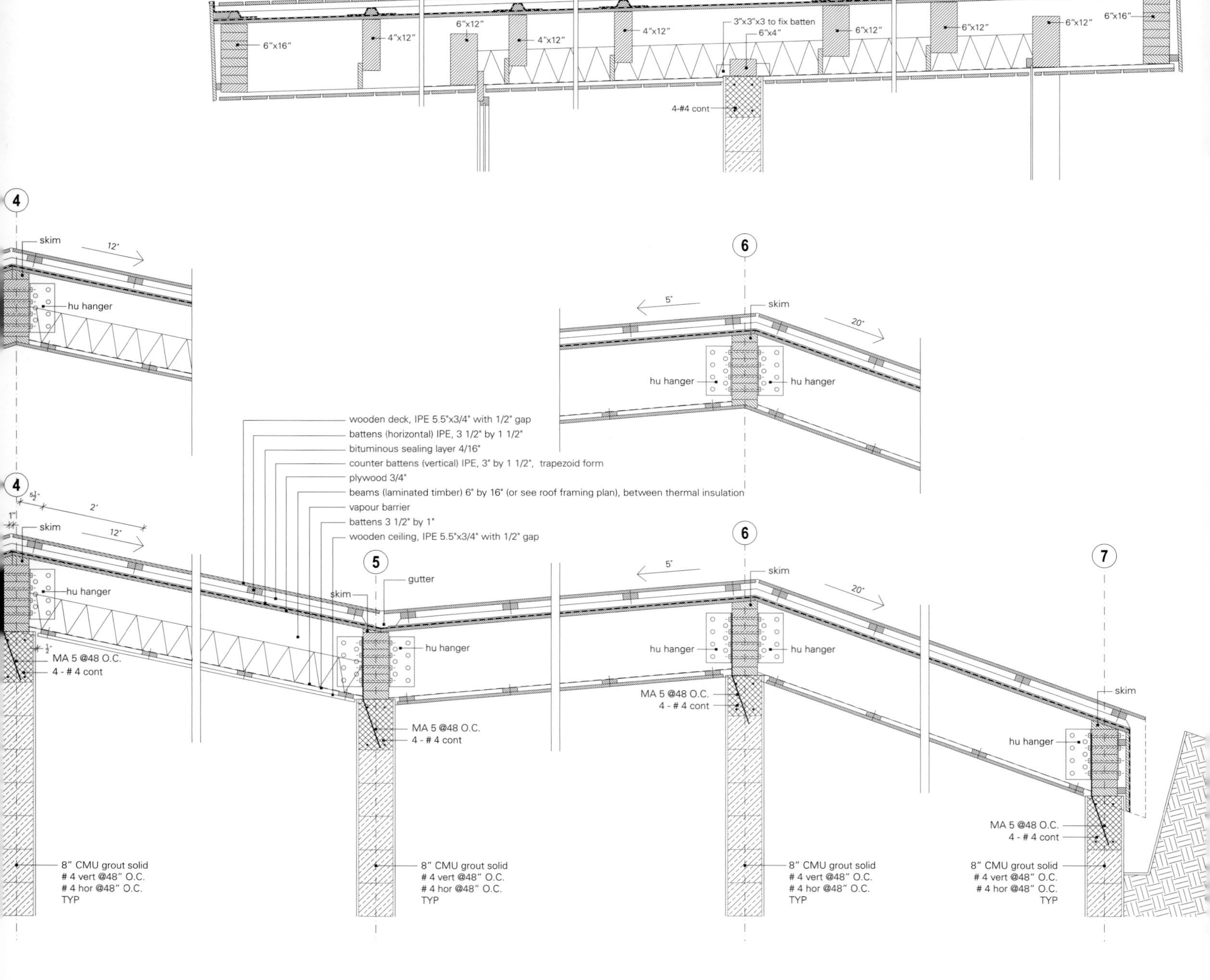

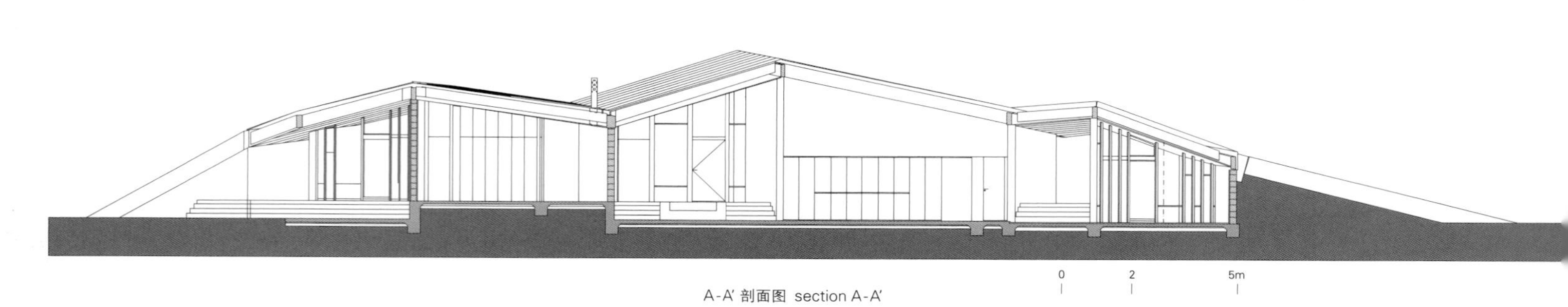

A-A' 剖面图 section A-A'

项目名称：Clifftop House Maui
地点：Maui, Hawaii, USA
建筑师：Dekleva Gregoric Arhitekti
项目团队：Aljoša Dekleva,Tina Gregorič, Flavio Coddou, Lea Kovič
甲方：Robert & Drazena Stroj
用地面积：10,000m² 建筑面积：450m²
总楼面面积：280m² 净楼面面积：263m²
项目设计时间：2004—2005
竣工时间：2011
摄影师：©Cristobal Palma (except as noted)

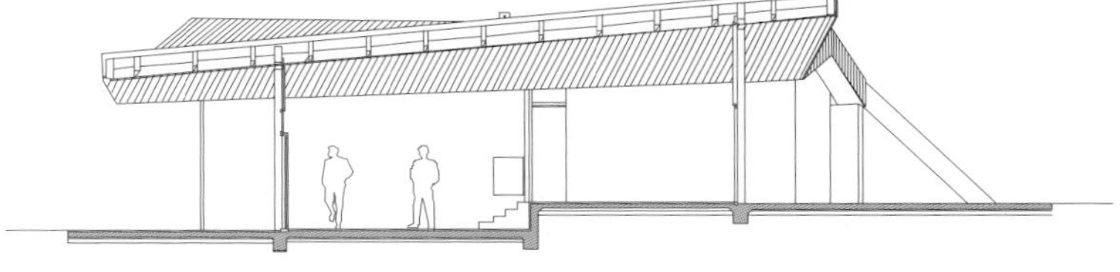

B-B′ 剖面图 section B-B′

Exit建筑师事务所

走近Exit

EXIT ARCHITECTS

Entering through the Exit

现代主义建筑被看作一场运动，在不止一个方面被证明是有害的。它的简单性转变为一种风格，这种简单性曾一度与城市的贫瘠以及以下层住宅高档化和都市的区域划分为基础的巨大破坏能力紧密相连，能够转变成一种对以复杂性和多样性为特征的城市环境采取介入手段的有益工具。Exit建筑师事务所的作品似乎以现代主义运动的几种本质特点为基础，包括以将其作品定位于大众为目的的强大媒体策略。他们的组织显然是开放式的，允许个人的努力以实用为目的相连接，无论何时（也无论怎样），只要是必需的，这给予这个事务所一种明显的当代形象。在严格的建筑关系中，该事务所的项目似乎从形式主义实验转向了类型学选择的简单化以及结构和技术探究的精妙深化。Exit建筑师事务所从小型居住空间的重整和室内装修任务开始，不断累积经验，从最开始只是竞标入围，到最后能够建成项目，这种缓和的策略在这个西班牙事务所的近期作品中似乎更加注重精炼的工艺和朴素无华的优雅性。

1.

自古，那些曾被描述成一场革命的运动看起来几乎都很适度，这些运动注定会解决由纪念性的、社会性的和生产性的转变所带来的危机和文化瓦解[1]所带来的挑战。然而之前的情况并非如此[2]。几十年来，那些一开始欲以解放建筑于客观性的诚实的尝试钻研于抽象、纯粹空间性和机械乌托邦形象的，其方式如此激进，其运作如此具有任意性，以至于人们无法想象它今日的温和，但是那些总是寻求自我推介和表面劝说的有效的宣传途径仍和19世纪20年代一样健康。议会、团体、先锋、杂志和宣言都已经简单地转移到网页和"网络"上，通过与最初的、最原始的明星建筑师相同的基本逻辑进行运营，即使在这个数字复制品时代也是如此[3]。

人们对于建筑在运营中作用的感知已经有所改变。人们的建筑方式以及对于真正的现代主义[4]本质的理解也已经通过一系列无法留存其最初的自大的事件而产生剧烈转变。现代主义建筑消逝于1972年3月16日15时[5]，或者是人们所被告知的类似时间；然而，断言它比从前更加健康，到处近乎狂野地遍地开花，也未为荒谬。没有了那些不能够理解其真正本质和范围（认为现代主义是一头失去控制的怪兽）的人所强加的巨大责任，现代主义建筑如今已经能够做回曾经的自己，更准确地说：仅仅是众多建筑风格中的一种。

Seen as a Movement, Modernist architecture proved noxious in more than one way. Turned into a style, its simplicity – once associated with urban sterility and an enormous destructive capacity based on gentrification and zoning – can be turned into an extremely useful tool for intervening within delicate urban contexts marked by complexity and multiplicity. The work of Exit Architects seems to feed on several essential traits of the Modern Movement, including a strong media strategy aimed at positioning their work in the public eye. Their organization, apparently open, allows for individual efforts to be joined pragmatically, whenever (and however) necessary, giving the form of this office an evident contemporary twist. In strict architectural terms, their projects seem to move from formalist experimentation to a simplification of typological choices and a deliberate deepening of configurational and technical inquiries. Stemming from the re-arrangement of small dwelling spaces and interior decoration commissions, and feeding of the experience gathered while moving from competition entries to buildable projects, the tempering strategy appears to favor refined craftsmanship and discreet elegance, in the most recent work of this Spanish collaboration.

1.

From a distance, what was once portrayed as a revolutionary Movement, destined to solve the crises posed by monumental social and productive transformations and the eminent risk of cultural disintegration[1], seems almost modest. This wasn't the case before, though[2]. For decades, what started as an honest attempt to liberate architecture from its object-hood, delving into the ether of abstraction, pure spatiality and images of mechanistic utopias, was so aggressive in its methods, so arbitrary in its operations, that one could hardly imagine its present-day meekness. But the effective propaganda methods, always seeking self-promotion and superficial persuasion, remain as healthy as they were in the nineteen-twenties. Congresses, groups, vanguards, magazines and manifestos have simply moved to web-pages and "networks", operating on the same fundamental logic of the first, original Starchitects, even in this age of digital reproduction[3].

The perception of architecture's role in the operation has changed. Our approach and understanding of the true nature of Modernism[4] has been crucially transformed by a series of events that made it impossible for its initial arrogance to persist. Modernist architecture died at 15:00, on March 16, 1972[5], or so we were told; and yet, it is not preposterous to affirm that it remains healthier than ever, sprouts blooming here and there, almost wild. Devoid of the titanic responsibilities given it by those who didn't understand its true nature and reach (turning it into a monster, literally out of control); it now has the possibility of being what it always was: simply

拆毁一座Pruitt-Igoe建筑，1972年4月
demolition of a Pruitt-Igoe building, April 1972

2.

现代主义运动的野心曾是这种风格所面对的众多问题之一。它态度孤傲，故意忽视地点而重视时间，实际上践踏了人类某些最宝贵的战利品。在一场完全缺乏常识的建筑秀中（到处都是这种建筑），甚至是来自反动政权的建筑革命分子也变成了工业世界的道德冠军，他们决定了所谓的好的和禁忌，同时带着明显的神祇般的偏见评估他人的堕落与卑鄙。

来自北部的白种人基督徒复制了他们祖先的模式，在拯救苍生的道路上通过他们神圣的介入手段解救世界于灾难之中。纯洁、纯粹主义、纯粹：干净的线脚，朴实无雕饰，自由的设计和开放的开口断面；一切都是理性的、机械的，充满技术性；解放于奇想、尘埃和黑暗之中。

设计竞赛通过创造受难的感觉完善了这种形象……现代主义建筑的先驱在不公的审讯中被击败或被罪恶扭曲的政治放逐：Iofan和Speer就像是Barabas，在满是Pilate[6]的世界中根本不支持那些被误解的英雄们。

但是最终，除所有的道德纯粹派之外，所有的宣传终被时间所淹没，所有的明星终都死去，他们的拥护者也几近逝去，所留下的却意味深长。无论它曾经多么成功，现代主义建筑的确曾导致灾难，它扼杀了城市（的多样性），同样扼杀了城市的居民，仅凭它干净完美、结实强健的本质就办到了。没有狄厄尼索斯（注：希腊神话中的酒神）参与其中，生活是不完整的。

3.

然而现代主义的遗物亦极其有趣。战后建筑在种种方面[7]进行改进，支持现代主义条理清晰的基本原则。[8]很明显，因为超越了嫁接到现代主义上的势不可挡的社会和政治责任，它也代表了未来在正式的试验方面的一次小进步，而同时又是一场在技术范围内作为建筑风格的大跨越。总的来说，绘画中对抽象的征服以及艺术中智力的主观性都允许一种开放的、复杂的艺术现象的构建：表面上风格薄弱的风格，又或者是一种接近于中立的风格。

是的，那些与古老建筑相关的联系（魔法师的把戏，方便地藏匿在一块桌布之后）现在非常清晰[9]，但在严格的创作关系中，基本在建筑结构领域之中，很明显，形式是可以追随功能的，至少在技术推动的、积极

a style – one among others, to make it even clearer.

2.

Among the many problems faced by the style was the Movement's ambition. Its autistic attitude, deliberately ignoring place in favor of time, literally trampled over some of humanity's most treasured conquests. In a show of total lack of common sense (full of themselves), architects-cum-revolutionaries, even from the most reactionary of political rights, turned into moral champions of the industrial world, deciding what was good and what was forbidden, evaluating others' decadence and baseness with a clear Apollinean bias.

White Christian men from the north, replicating the model of their forefathers, pontificated on the road to salvation, saving the world from catastrophe through (their) almost divine intervention. Purity, purism, pureness: clean lines, no ornament, free plans and open sections; everything rational, mechanical, technical; liberated from whim, from dirt, from darkness.

Competitions completed the figure, by creating the sensation of martyrdom... a modern pioneer being defeated in unfair trial or being politically ostracized in favor of evil: Iofan and Speer as easy Barabas, opposed to misunderstood heroes in a world full of Pilates[6].

But in the end, all moralist purism aside, all propaganda flattened by time, all stars dead and their groupies about to pass away, what remains is eloquent. Whenever truly successful, Modernist architecture led to disaster, killing cities and inhabitants alike, simply by being so perfectly clean, so naturally tense and muscular. Without Dionysos in the operation, life is incomplete.

3.

What remains is extremely interesting, though. Postwar architecture has operated on myriad directions[7], supported on the formal basis of Modernism.[8] Because it is clear that, way beyond the overwhelming social and political responsibilities grafted onto it, Modernist architecture represented a small movement forward in terms of formal experimentation, and yet a big move upward in the scale of technique as architectural style. The conquest of abstraction in painting, and intellectual subjectivity in art as a whole, allowed for the construction of an open and complex artistic phenomenon: the apparently styleless style, or something very close to neutrality.

Yes, links with older architectures (magicians' tricks, conveniently hidden behind a tablecloth) are clear now[9]; but in strict compositional terms, fundamentally in the realm of architectural configuration, it is evident that, at least in the positive heuristic field of technical motivation (firmitas), form can follow function, as opposed to the illusion of activity-defined typologies.

The arrogant cry of the self-designated pioneers has turned into a beautiful word: simplicity (which some now confuse for "minimal-

1 *All That is Solid Melts into Air: The Experience of Modernity*, Marshall Berman, New York: Penguin Books, 1982
2 *Programs and Manifestoes on 20th-century Architecture*, Ulrich Conrads, Cambridge: MIT Press, 1970
3 "The Work of Art In the Age of Mechanical Reproduction", *Illuminations*, Walter Benjamin, London: Fontana Press, 1972
4 "简单地说，现代化是指社会在技术和社会经济范畴内的创新过程。现代化意味着这一过程的实践，以及技术和社会经济变化过程中所产生的条件。最终，现代化代表着这一条件下的艺术和智能反应。简言之，就是在艺术和文化上对其自身的体现。" *Architectural Positions: Architecture, Modernity and the Public Sphere*, Tom Avermaete, et al. (Eds.), Amsterdam: Sun Publishers, 2010, pg. 19. This brief text submits to: Hilde Heynen, *Architecture and Modernity: A Critique*, Cambridge: MIT Press, 1999, p. 26
5 暗指Minoury Yamasaki的St. Louis Pruitt-Igoe项目的拆毁。*The Language of Postmodern Architecture*, Charles Jencks, New York: Rizzoli, 1984
6 1927年国际联盟日内瓦总部万国宫设计竞赛的评委会。与柯布西耶的现代主义提案远远不同，他们选择组织一个由决赛选手（Broggi、Flegenheimer、Lefevre、Nenot、Vago）组成的团队，采用一种不同的风格来设计项目。*Anxious Modernisms: Experimentation in Postwar Architectural Culture*, Sarah

1. *All That is Solid Melts into Air: The Experience of Modernity*, Marshall Berman, New York: Penguin Books, 1982
2. *Programs and Manifestoes on 20th-century Architecture*, Ulrich Conrads, Cambridge: MIT Press, 1970
3. "The Work of Art In the Age of Mechanical Reproduction", *Illuminations*, Walter Benjamin, London: Fontana Press, 1972
4. "In brief, 'modernization' refers to the process of innovation in society in the technical and socio-economic spheres. 'Modernity' signifies the experience of this process, the condition arising from these processes of technical and socio-economic change. Finally, 'modernism' stands for artistic and intellectual reflection on this condition – in short, the way that this manifests itself in art and culture.", *Architectural Positions: Architecture, Modernity and the Public Sphere*, Tom Avermaete, et al. (Eds.), Amsterdam: Sun Publishers, 2010, pg. 19. This brief text submits to: Hilde Heynen, *Architecture and Modernity: A Critique*, Cambridge: MIT Press, 1999, p. 26
5. Alluding to the demolition of Minoury Yamasaki's St. Louis Pruitt-Igoe project. *The Language of Post-modern Architecture*, Charles Jencks, New York: Rizzoli, 1984
6. The Jury of the 1927 Palais de Nations competition for the Geneva Headquarters of the League of Nations comes to mind. Far from Corbusier's modernist proposal, they chose to assemble a team of finalists (Broggi, Flegenheimer, Lefevre, Nenot and Vago) to design a project with a different style.

的探索领域中是这样，而不是幻想用行为定义类型。那些自我认定的先锋们自大的呼喊已经变为一个优美的词语：简约（该词现在常常与"极简派艺术"一词相混淆）。所谓的革命转变为改良，纯粹的心灵暴力变为无法被认出的驯服。作为一种风格，现代主义建筑如此谨慎、如此缄默，以至于它似乎成了修补和完善它从前的敌人——传统城市——的完美工具。

在文化方面经过了数十年的反复试验之后，这个星球上的每一个小城镇都已经变成了一块紧张之地，不同风格的建筑并存，现代主义也毫不例外。然而，建筑风格也不得不做出让步：从卡萨布兰卡的城市街区到伦敦遗留的小巷（现在被极小的日式盒式房屋所占据），从加拿大的郊区到拉丁美洲的广场，Domino和Citrohan类型已经设法重建乡村并通过展览馆来运营（有时甚至是同时的）。只要它们的时代一到来，它们便学习文明的人类所学习去做的：和平地、互相尊敬地彼此共生，却又不失去它们各自的特性。

有人称之为成熟。

4.

位于马德里的Exit建筑师事务所最有趣的一些作品刚好与用现代主义进行全新释义的计划完美匹配，计划中实用主义已经通过新的媒体传播，竞争激烈，现代主义风格的建筑非常实用。Exit建筑师事务所通过"五人行动"[10]这一开放式的计划施行，该计划十分灵活，就如同《先锋》杂志的成员们在20世纪初所能达到的灵活程度一般。当达达主义者或超现实主义者很少束缚于一本杂志的时候，一场展览会或一个咖啡馆便能转变成政府总部；当代建筑师们既有个人行动也有集体行动，既运作研究计划也外包一切他们能够外包出去的，这一切总是被网络的复杂性所支持。

Exit建筑师事务所的策略是很明显的：今日的工作室由引人注目的"协作者"团体[11]、技术建筑师和顾问所支撑，因此工作室便能创作出足够的"临界物质"，成功地提交参赛作品参加大量的设计竞赛。接着便有了柯布西耶媒体技巧：除去114页的巨幅下载目录，网页中包含每一名成员的工程、建筑、公共演讲和出版物；除此之外，类似于Jeanneret出版商及其助手的PR工具已经设法在网络空间将完全一样的信息（平面图、剖面图、照片、描述）传播开来——这意味着这些将无所不在！从Exit建筑师事务所自己的竣工建筑作品展到当代西班牙建筑的集体展，从微博到杂志再到书籍（包括本书），Exit建筑师事务所清楚地意识到，当他们

ism"). What was proposed as revolutionary turns into reform, and the violence of the tabula rasa becomes unrecognizably tame. As a style, Modernist architecture is so discreet and silent, that it appears to be the perfect tool for mending and completing its former nemesis: the traditional city.

After decades of cultural trial, every town in the planet has become a field of tension where different architectures coexist, and Modernism isn't an exception. It has had to make concessions though: from the urban blocks of Casablanca to the tiny leftover alleys of London (now being occupied Japanese-style with infinitesimal box houses); from the Canadian suburb to the Latin American plazas, the Domino and the Citrohan typologies have managed to recreate courtyards and operate as pavilions (sometimes even simultaneously), learning what civilized people learn to do, as soon as they come of age: coexisting peacefully, respectfully, without losing their own individuality.

Some call this maturity.

4.

Some of the most interesting work by Madrid based collective Exit Architects fits perfectly within the scheme of this new interpretation of Modernism, in which pragmatism has been boosted by new media, competition is fierce, and the Modernist style is very useful. A five-man-act[10], Exit operates on an open scheme, as flexible as the members of a Vanguard could be at the dawn of the 20th century. While Dadaists or Surrealists were loosely tied by a magazine, an exhibition or a café turned official headquarters; contemporary architects have individual and collective acts, operate on research programs and outsource whatever they can, always supported by the complexity of the Web.

Their strategy is clear: Supported by an impressive body of "collaborators"[11], technical architects and consultants, present-day ateliers create enough "critical mass" to successfully submit entries to a great deal of competitions. And then there's the Corbusiar media trick: aside from the massive 114-page downloadable catalog, the web-page including each of the members' projects, buildings, public talks and publications; aside from all this, a PR apparatus akin to Jeanneret's publishers and acolytes has managed to post the exact same information (plans, sections, photographs, descriptions) in cyberspace – meaning everywhere! From exhibitions within their own finished buildings to collective shows of contemporary Spanish architecture; from blogs to magazines to books (this one included); Exit Architecture definitely understands that when it comes to building a name for themselves, Mies's famous dictum[12] is definitely passé.

Focusing on their architecture, it is possible to identify a movement from quantity to quality, and from the ether of impacting

Williams Goldhagen and Réjean Legault (Eds.), Cambridge (Mass.): MIT, 2001
7 *The Formal Basis of Modern Architecture*, Peter Eisenman, Baden: Lars Müller Publishers, 2006
8 *Essays in Architectural Criticism: Modern Architecture and Historical Change*, Alan Colquhoun, Cambridge: MIT Press, 1981.
9 *The Mathematics of the Ideal Villa and Other Essays*, Colin Rowe, Cambridge: MIT Press, 1982
10 建筑师Silvia N. Gómez作为公司的合伙人之一出现在下载目录里，她在最近的在线网页中被建筑师José María Tabuyo代替（http://exit-architects.com/01%20ESTUDIO/exit-peril.html——2012年5月19日检索）。
11 现今的从业者或学徒工，即通常我们所知的实习生，使公司在经济上实现人力资源独立。依靠免费的或是十分廉价的劳动力，使无需真正支付相应的佣金就能完成大量工作成为可能。
12 "少即是多"

Anxious Modernisms: Experimentation in Postwar Architectural Culture, Sarah Williams Goldhagen and Réjean Legault (Eds.), Cambridge (Mass.): MIT, 2001
7. *The Formal Basis of Modern Architecture*, Peter Eisenman, Baden: Lars Müller Publishers, 2006
8. *Essays in Architectural Criticism: Modern Architecture and Historical Change*, Alan Colquhoun, Cambridge: MIT Press, 1981.
9. *The Mathematics of the Ideal Villa and Other Essays*, Colin Rowe, Cambridge: MIT Press, 1982
10. While architect Silvia N. Gómez appears in the downloadable catalog of the firm as a Partner, she has been replaced by architect José María Tabuyo in the current online version of the webpage (http://exit-architects.com/01%20ESTUDIO/exit-perfil.html - retrieved on May 19, 2012).
11. The current figure of the practitioner or apprentice, usually known as "Intern", allows for firms to count on manpower independent of economy. Counting on free or extremely cheap labor, it is possible to produce an enormous volume of work without the need to land true commissions to pay for it.
12. "Less is more"

开始为自己创名时，密斯的著名箴言[12]就已经远去了。

关注Exit建筑师事务所设计的建筑，可以看到一场从数量到质量、从影响形象的设计到只有真正的试验才能提供的可靠性的运动。即使是在他们自我推介的介绍中，同行们也能将他们的方法搞清。从客户的满意度到高质量的建筑再到有效的流程管理，他们的目标完全切合实际。

他们的抱负、竞争的睿智使其意图变得模糊，至少是部分意图。从城市设计到室内装饰，包括景观、社会住房和公共纪念建筑，虽然Exit建筑师事务所从事各种设计，不过看似涉及的范围还过于狭窄，或者在某一段时间内如此。但是在欧洲项目中不怎么幸运的尝试，或者内华达塞拉利昂游客中心异想天开的形式主义却通过一系列小的委托任务得到了弥补，这些项目包括马德里小住宅和公寓的重建、一家餐馆的装饰，这些项目都是在建筑工艺上的真正实践。

Exit建筑师事务所更加温和、成熟，似乎进入了相当保守而又简单（因此安全）的类型学试验领域，带着之前提到的现代主义风格的优雅进行构建。从马德里Las Rozas学校的亭子体系，到同样在马德里市的Ofelia Nieto公寓的Coderch式外立面，都体现出建筑正缓慢但却坚定地走向简单化。在获奖的参赛作品和实际的建筑工程之间总是存在着大量的变动，这被理解为真正的现实检验。Exit建筑师事务所连接了这一沟壑，现在正踏着严肃强健的建筑工艺之路前行。如此，他们便提供给所有人值得沉思的素材。

5.

如前所述，自从20世纪50年代末第一次尝试去颠覆和超越霸权运动开始，人们就写作了大量文章不断求新。每一个修复主义想法的较小（大多数情况为天真）的目的都是幻想出一种新建筑，和这场运动一样"彻底"，但鉴于它与一两种积极的探索相抵触，因此又完全不同，然而它又通过完全相同的基本原理进行运作。

像后现代和解构主义——或因此而试图使自身成为"建筑"的现在的思想突变——这样极其悲哀的烙印，已成为这一策略失败的力证。通过使格罗皮乌斯和海伯森默的建筑具有真正的风格，并使其写入撰写城市人类历史的专业人士的作品之中，Exit建筑师事务所似乎已经认识到现代主义风格能够成为一种有益工具。因此，他们为人们开启了一扇门；人们亦能进入。

images to the solidity that only true experience can provide. Even in their self-promotional introduction, our colleagues make their no-nonsense approach clear. Their aims are totally down-to-earth, ranging from customer satisfaction to high quality construction and efficient process management.

The breadth of their ambition, competition-wisdom, blur the intention, at least partially. By tackling chores ranging from urban design to interior decoration, including landscaping and social housing and public monuments, Exit appears to spread itself too thin, or appears to do so for some time. But the less fortunate attempts of the Europan projects, or the rather whimsical formalism of the Sierra Nevada Visitor Centre appear to be tempered by a set of tiny commissions, including the re-organization of small houses and flats in Madrid, and the decoration of a restaurant, all true exercises in architectural craftsmanship.

More tempered and mature, Exit Architects appears to move into the realm of rather conservative and simple (and therefore safe) typological experiments, configured with the elegance of the aforementioned Modernist style. From the pavilion-system of the Las Rozas school, in Madrid, to the Coderch-like facade of the Ofelia Nieto apartments in the same capital, there appears to be a slow but strong movement towards simplicity. Between prized competition entries and the actually built projects there are always substantial changes, understood as true reality-checks. Exit Architects has bridged this gap, and now tread the path of sober and robust architectural craftsmanship. In doing so, they provide all of us with material for reflection.

5.

As mentioned, lots of writing have been done, since the first attempts to subvert and overcome the hegemonic movement in the late fifties, always searching for the new. Every revisionist intention has met the lighter (and in most cases naïve) aims to create the illusion of a new architecture, as "complete" as that of the Movement, but totally different, in the sense of competing against one or two of its positive heuristics; and yet operating on the exact same rationale. Extremely sad brands, such as eclectic post-mod and deconstructivism – or for that sake, current boutades attempting to become "architectures" in their own right – are a clear show of the failure of this strategy. By bringing the architecture of Gropius and Hilberseimer to its true stylistic dimension, though, and applying it to the patient exercise of the professional who writes infinitesimal portions of the history of humanity in the palimpsest of the city, Exit Architects appear to have realized that the Modernist style can be a very useful tool. Thus, they open a door for us; and we can also enter. Jorge Alberto Mejía Hernández

帕伦西亚监狱改造的文娱中心

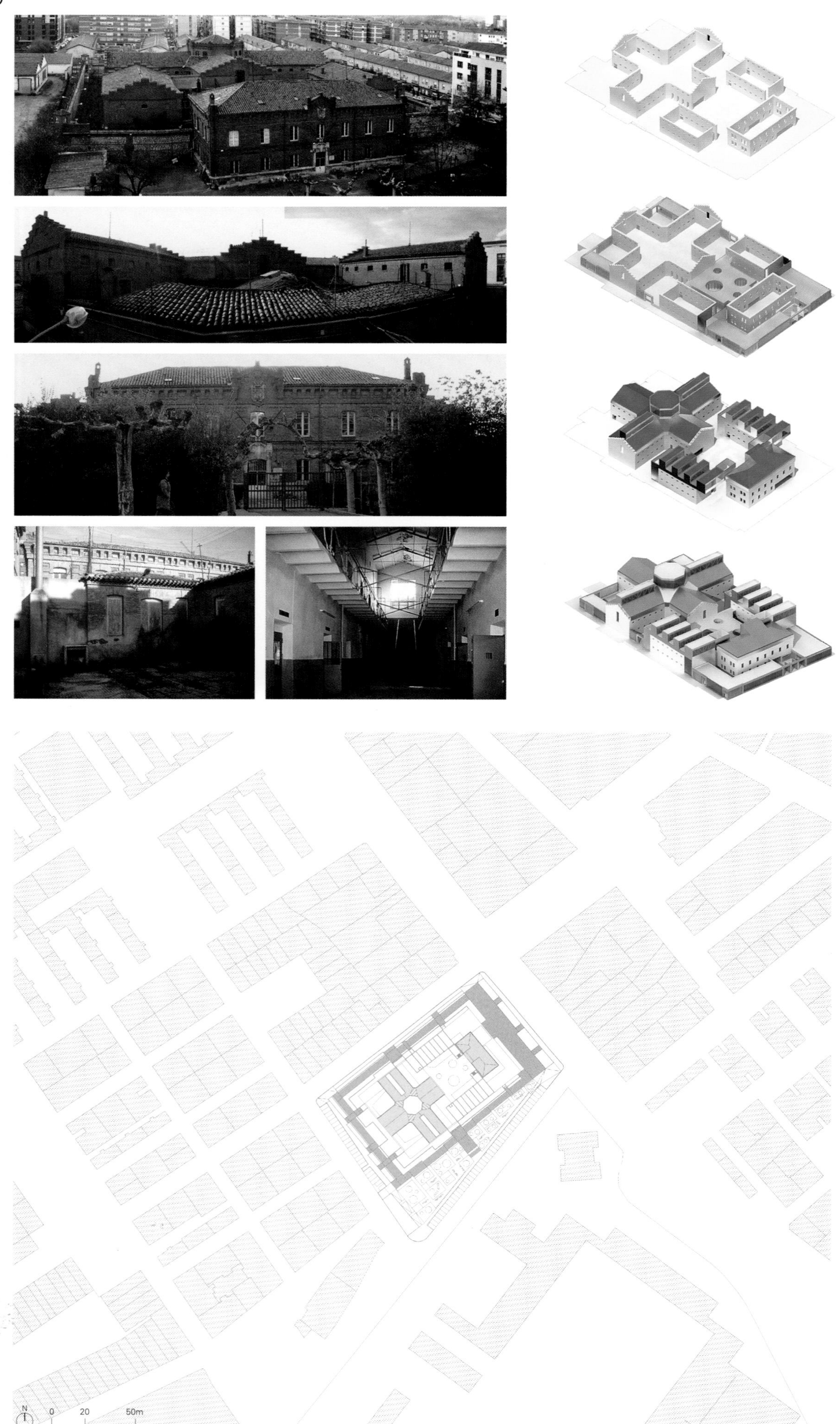
N
0
20
50m

改造老旧建筑通常需要全面规划，全方位考虑，需要同时经营管理建筑工程的多个领域。无论多么有才华的专业人员，在绘图时也可能会因为只是一味地考虑某一单个问题而忽视了工程的关键地方。

当参照根深蒂固的历史和理论方面的准则原理时，感性会创造奇迹。但如果基础被忽视，工程看起来就会不完整，最好的情况也是不那么令人满意。

陈旧偏远的帕伦西亚监狱（西班牙卡斯蒂利亚–莱昂自治区）的改造工程项目竞争激烈，说明该项目运作过程要求严格正式。该项目是对要求很高的圆形监狱辐射状建筑类型进行改建。砖石结构的圆形监狱建于19世纪，由一个简单的四翼结构构成，中间通过一个八角形灯笼式天窗连接，占据矩形地块的一侧。面对它时能看到这是一座U形的建筑，封闭了地块的短边，这里也是整个监狱的中央庭院。

Exit建筑师事务所倾向于通过显著增加建筑密度以使整体更加"完整"的方法来对圆形监狱进行改建。改建后的圆形监狱，从里面看似乎只是对旧板材和屋顶材料进行了简单更换，其实在形态学上是有极大改变的。改建后，每个展馆的中央走廊都敞开了，通过上空空间可以俯瞰到一层。

在U形建筑北部较矮的侧翼结构的一层有一个礼堂和几间教室，上面一层有两间多功能展厅，高度与主建筑相同，这就使本来就很复杂的屋顶设计变得更加复杂。U形建筑较长的中间区域是办公室，使原本必须对称的设计转变成了传统的解决方案，这样做似乎可以使人在正式严谨的工作中得到放松，也有利于使建筑功能简洁化。

对已建成的完整建筑进行改造，虽然是上空空间的应用使效果更显著，但与其元素组成有极大关系。在这方面，Exit建筑师事务所赞成沙利文的观点，认为应该在建筑功能方面多下工夫而不是外形。中央庭院被完全覆盖，它的威严性也被SANAA风格的建筑取而代之，即在此创造性地设置了圆形采光井（当然了，不像洛桑垫那样倾斜）；外围有好多浴室、自助餐厅以及大量的服务空间，这些空间显然不会影响建筑的主要形状。因此，就像在前文中介绍的那样，该项目是一座与其他建筑成功叠加的中性的光滑漂亮上相的建筑，是一座不算华丽（至少在外观方面）的建筑，必定曾经是一座监狱。

至少在这种情况下，即使组成策略看起来不充足，即使优雅性因工程及其发展潜力的基本形态学的明显误读而降低，即使是一个明显缺乏同情心、平凡的、粗野的、爱在暗里地搞破坏的老家伙也可能给人带来更舒心的礼物，而不是华而不实的东西。

Palencia Prison Renovation to Cultural Civic Centre

Interventions on older buildings usually require a multi-layered approach, operating on several fields of architectural activity simultaneously. However talented, it is not unusual for professionals to focus on a single problematic aspect of the task they face, missing the crux of what lays on their drafting boards.

When paired with a deeply rooted understanding of the fundamental historical and theoretical aspects of the discipline, sensibility works wonders. When the base is overlooked, though, projects appear incomplete, or not that satisfactory, at best.

The prized competition entry for the refurbishment of the old provincial prison of Palencia (autonomous community of Castilla y León, Spain) implies a delicate formal operation. Reproducing the exceedingly demanding radiated typology of the Panopticon, the original 19th century masonry building consisted of a simple four-arm cross, articulated by a central octagonal lantern, occupy-

项目名称：Rehabilitation of former prison of Palencia as cultural civic centre
地点：Palencia, Spain
建筑师：Ángel sevillano, José Mª Tabuyo
项目团队：Mario sanjuán, Ibán Carpintero, Miguel García-Redondo
工料测量员：Impulso Industrial Alternativo. Álvaro Fernández
结构工程师：NB35. José Luis Lucero
机械工程师：Grupo JG. Juan Antonio Posadas
照明顾问：Manuel Díaz Carretero
总承包商：COPISA CONSTRUCTORA PIRENAICA S.A.
甲方：Ministerio de Fomento / Ayuntamiento de Palencia
用地面积：8,258m²
建筑面积：5,077m²
总楼面面积：3,058m²
造价：EUR 9,675,038
设计时间：2007
竣工时间：2011
摄影师：© FG+SG Architectural Photography

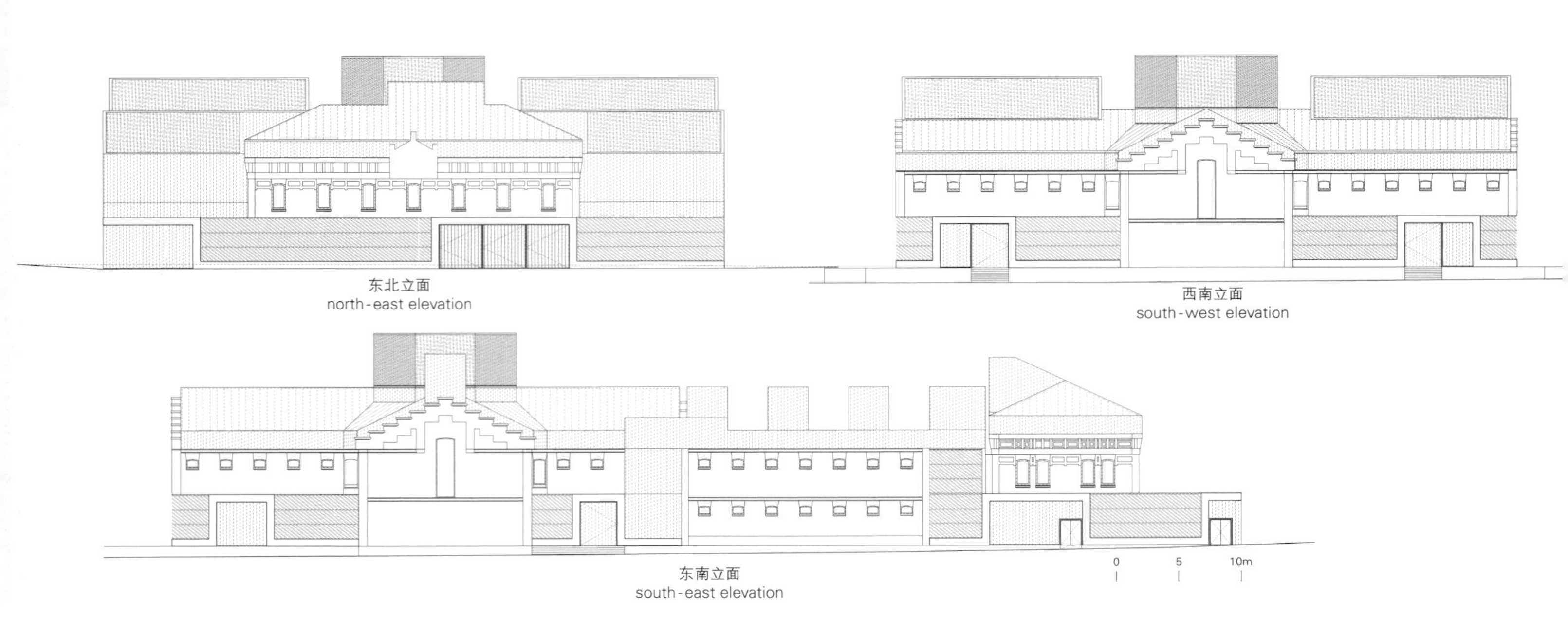

东北立面
north-east elevation

西南立面
south-west elevation

东南立面
south-east elevation

ing one side of the rectangular plot. Facing it, a U-shaped block closed the perimeter of the plot on its shortest side, forming a central courtyard for the whole facility.

Exit Architects' intervention has focused on "completing" the figure, by increasing the building's density dramatically. Work inside the Panopticon, presented as a simple replacement of old slabs and roof materials, is actually a radical morphological inversion, that leaves the central corridors of each pavilion open, void space overlooking the ground floor.

Atop the lower, lateral arms of the northern U, colonized on their ground floor with an auditorium and classrooms, two multi-purpose exhibition halls have risen to the height of the main cross, while introducing complexity in their intricate roof design. The longer, central side of the U is used for offices, turning what must have been symmetrical into a conventional solution that appears to relax on formal rigor in favor of functional simplicity.

Those, the operations performed on full, built blocks. What becomes more striking, though, is the use of void space, highly magnetized by the power of such an elemental composition. In this respect, Exit appeals to Sullivan's misquote, tackling the functional aspect rather than paying justice to the shapes they deal with. The central courtyard is totally covered, its majesty replaced by SANAA-like, whimsically placed circular light wells (without the sloping of the Lausanne mat, of course); the perimeter literally stuffed with bathrooms and cafeterias and a myriad servant spaces that apparently didn't make it to the main shape. The result, as mentioned in the introduction to these projects, is a slick, photogenic building that successfully superimposes a neutral style on another, less gorgeous (at least in terms of its outside face) edifice: a prison, no doubt.

At least in this case, though, this make up strategy appears insufficient; elegance muffled by an apparent misreading of the fundamental morphology of the project and its possibilities; discreteness undermined by an apparent lack of empathy with an old fellow who, however prosaic and tough, could have been brought to the present more comfortably, instead of overdressed.

Jorge Alberto Mejía Hernández

A-A′ 剖面图 section A-A′

B-B′ 剖面图 section B-B′

C-C′ 剖面图 section C-C′

D-D′ 剖面图 section D-D′

E-E′ 剖面图 section E-E′

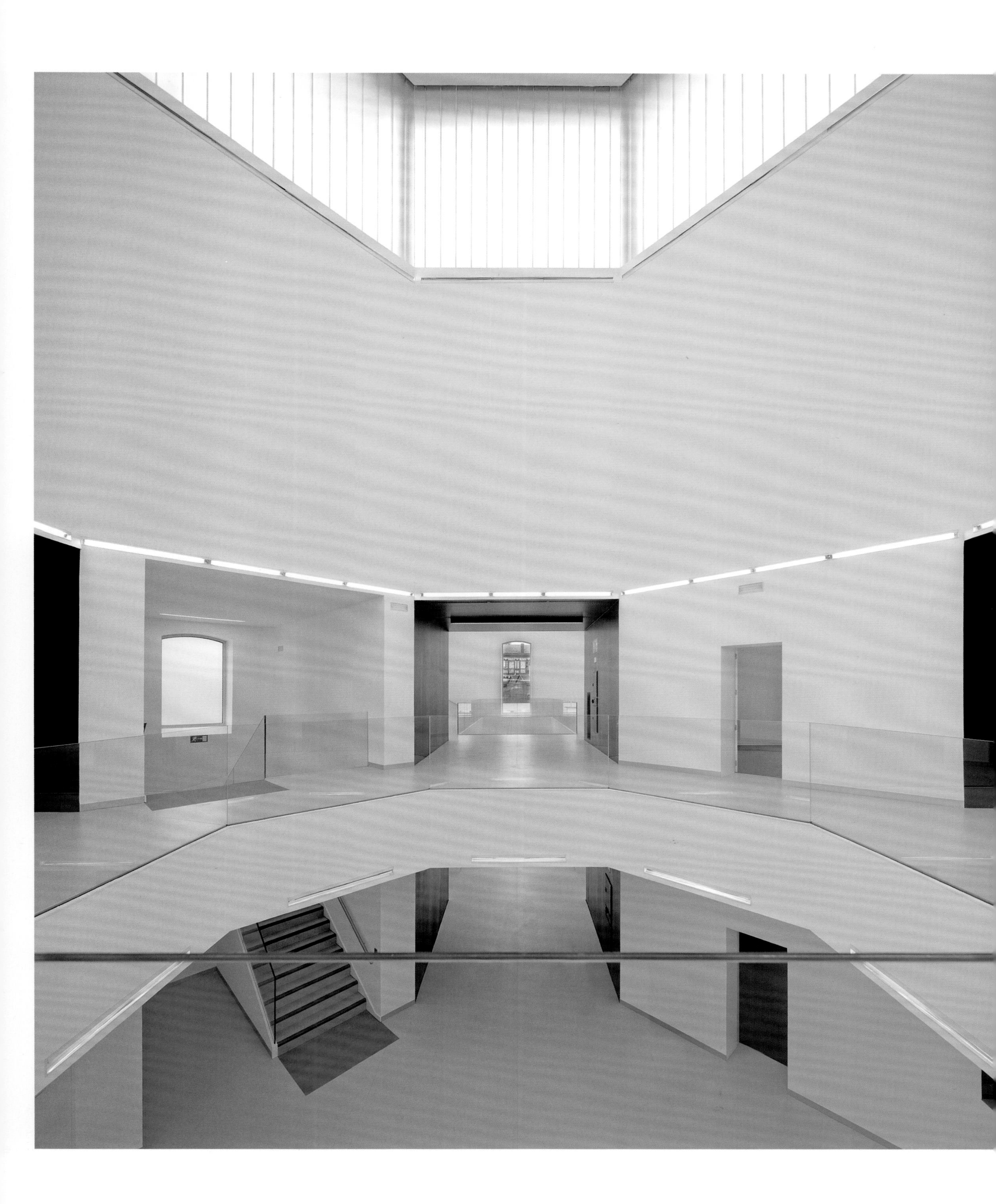

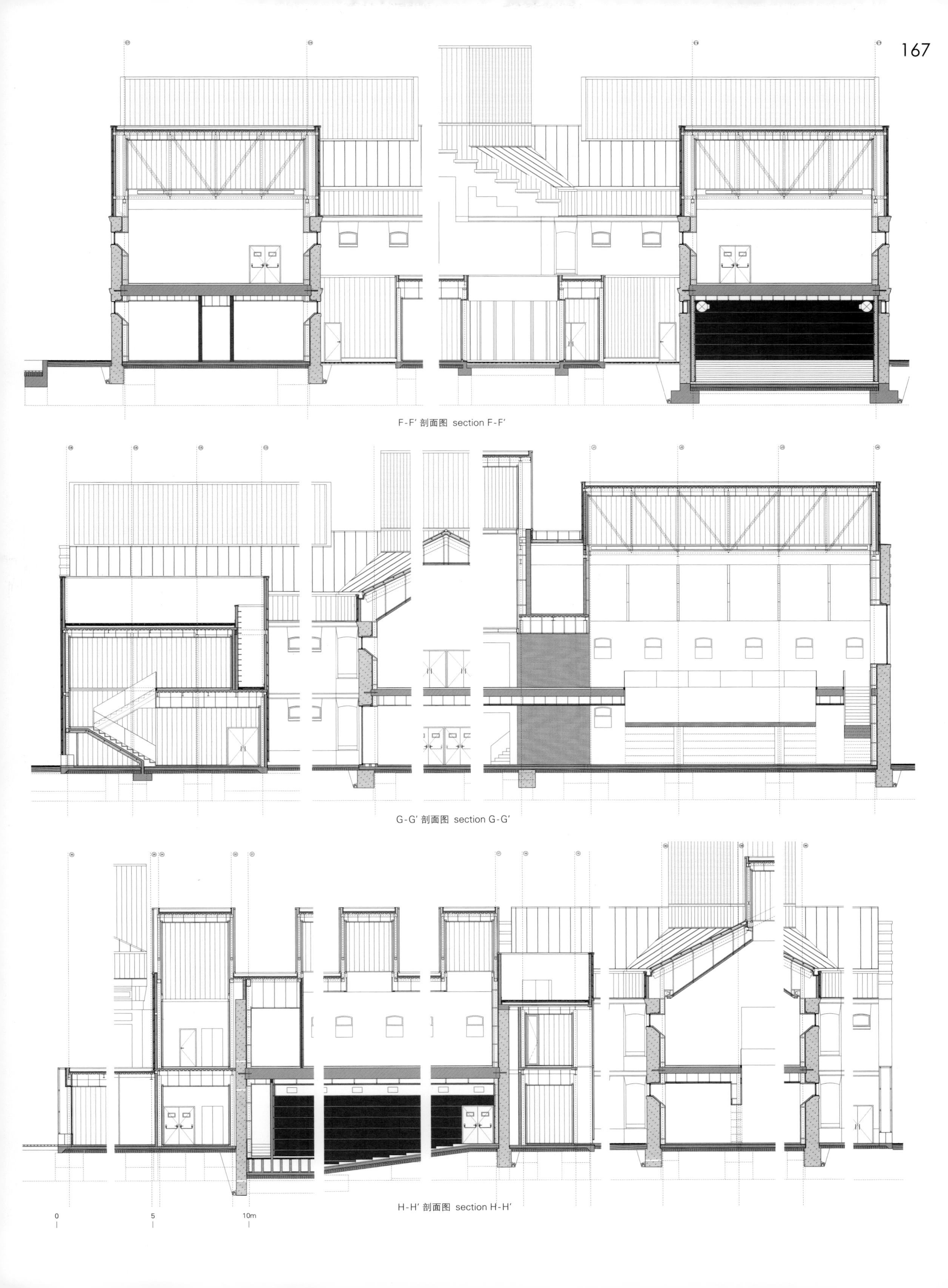

F-F′ 剖面图 section F-F′

G-G′ 剖面图 section G-G′

H-H′ 剖面图 section H-H′

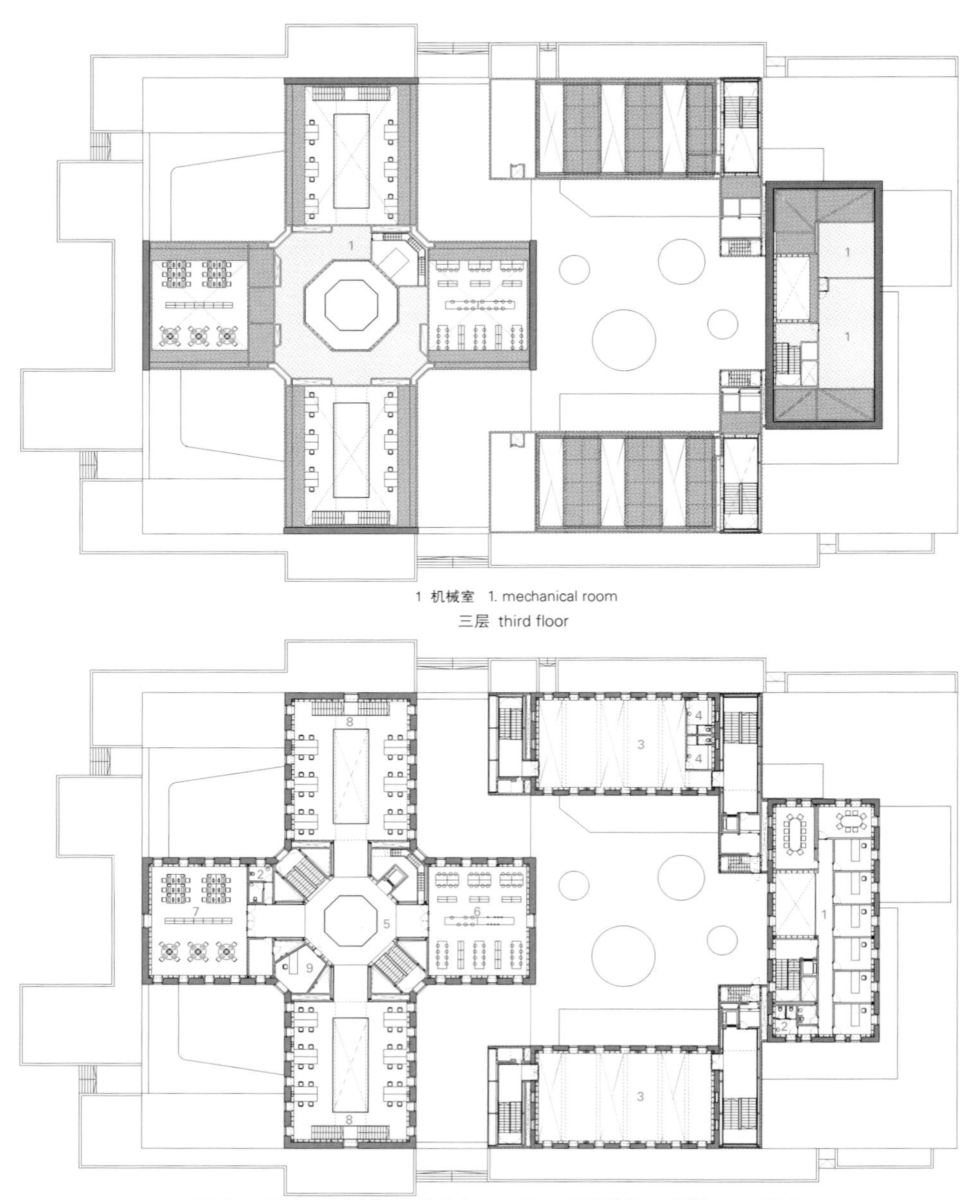

1 机械室　1. mechanical room

三层　third floor

1 办公室　2 卫生间　3 多功能区　4 更衣室　5 走廊　6 儿童阅览室　7 多媒体室　8 阅览室

1. offices 2. toilets 3. multi-purpose areas 4. dressing room 5. corridor 6. children reading room 7. multi-media room 8. reading room

二层　second floor

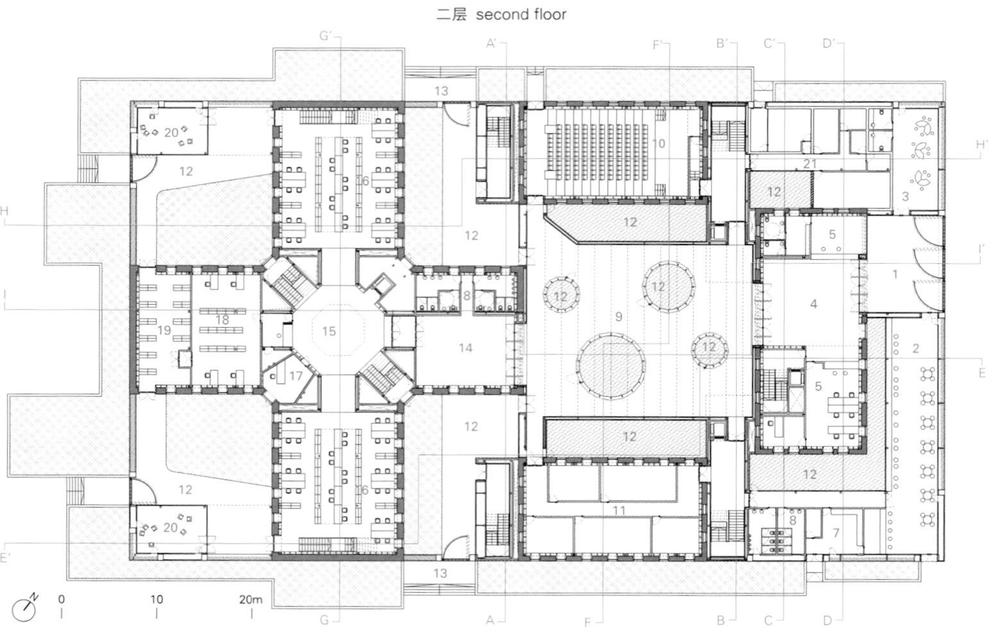

1 主入口　2 自助餐厅　3 托儿所　4 接待处　5 控制室　6 办公室　7 厨房　8 卫生间　9 大厅和展览画廊　10 大礼堂　11 音乐及艺术教室　12 庭院　13 次入口　14 图书馆大厅　15 图书馆接待处　16 阅览室　17 图书档案室　18 图书储存室　19 休息室　20 设备间

1. main access 2. cafeteria 3. nursery 4. reception 5. control 6. offices 7. kitchen 8. toilets 9. hall & exhibitions gallery 10. auditorium 11. music & art classroom 12. courtyard 13. secondary access 14. library hall 15. library reception 16. reading room 17. book archive 18. book storage 19. rest room 20. installations

一层　first floor

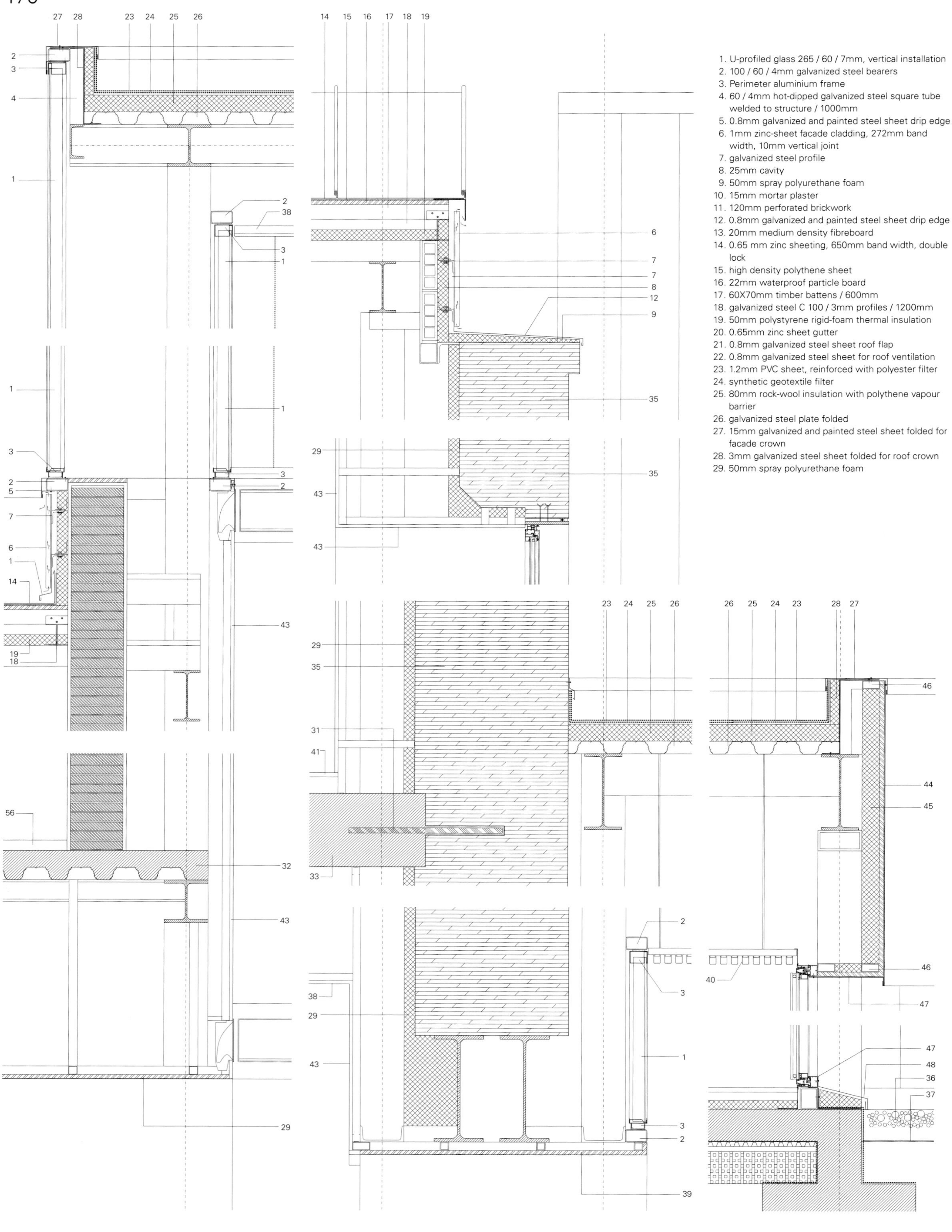
1. U-profiled glass 265 / 60 / 7mm, vertical installation
2. 100 / 60 / 4mm galvanized steel bearers
3. Perimeter aluminium frame
4. 60 / 4mm hot-dipped galvanized steel square tube welded to structure / 1000mm
5. 0.8mm galvanized and painted steel sheet drip edge
6. 1mm zinc-sheet facade cladding, 272mm band width, 10mm vertical joint
7. galvanized steel profile
8. 25mm cavity
9. 50mm spray polyurethane foam
10. 15mm mortar plaster
11. 120mm perforated brickwork
12. 0.8mm galvanized and painted steel sheet drip edge
13. 20mm medium density fibreboard
14. 0.65 mm zinc sheeting, 650mm band width, double lock
15. high density polythene sheet
16. 22mm waterproof particle board
17. 60X70mm timber battens / 600mm
18. galvanized steel C 100 / 3mm profiles / 1200mm
19. 50mm polystyrene rigid-foam thermal insulation
20. 0.65mm zinc sheet gutter
21. 0.8mm galvanized steel sheet roof flap
22. 0.8mm galvanized steel sheet for roof ventilation
23. 1.2mm PVC sheet, reinforced with polyester filter
24. synthetic geotextile filter
25. 80mm rock-wool insulation with polythene vapour barrier
26. galvanized steel plate folded
27. 15mm galvanized and painted steel sheet folded for facade crown
28. 3mm galvanized steel sheet folded for roof crown
29. 50mm spray polyurethane foam

30. 2mm hot-dipped galvanized steel louvres, welded to steel frame
31. ø25mm drilled hole, filled with thixotropic - epoxy resin and steel rebar / 500mm, for connection to existing brick wall
32. reinforced concrete composite floor slab
33. reinforced concrete waffle slab
34. 550mm existing brick wall to be rehabilitated
35. 700mm existing brick wall to be rehabilitated
36. 150 mm selected gravel 80 / 100mm
37. synthetic geotextile filter
38. plasterboard ceiling
39. steel sheet ceiling
40. 40 / 25mm U-profiled aluminium ceiling
41. acoustic PVC flooring
42. high traffic PVC flooring
43. 2X15 mm plasterboard with paint finish
44. 0.6 mm galvanized and painted steel sheeting, over 19mm medium density fibreboard
45. 80mm polystyrene rigid-foam thermal insulation
46. 80 / 40 / 4mm hot-dipped galvanized steel square tube
47. aluminum joineries
48. 0.8mm galvanized and painted steel sheet folded for rain guard
49. 150mm concrete flooring
50. drainage polyethylene sheet waterproofing
51. sand bed
52. 150mm gravel 40 / 80
53. 40mm polystyrene rigid-foam insulation
54. 150mm continuous washed concrete flooring, white cement
55. 150mm artificial gravel
56. washed concrete slab

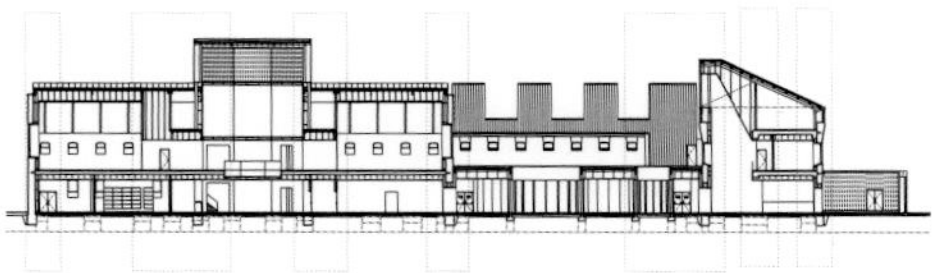

I-I' 剖面图 section I-I'

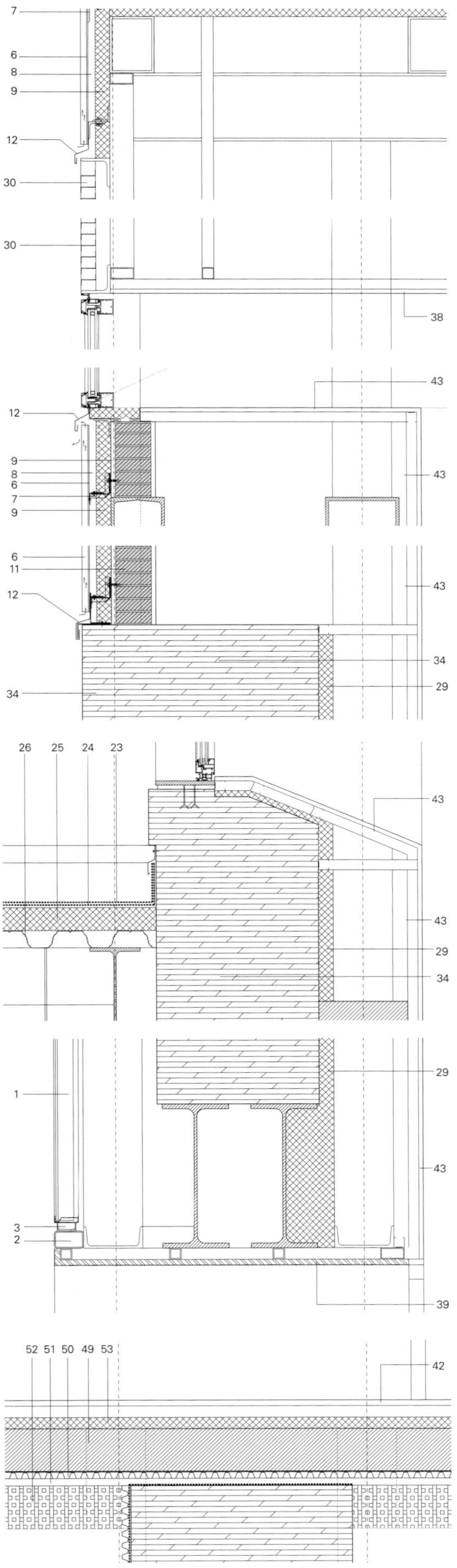

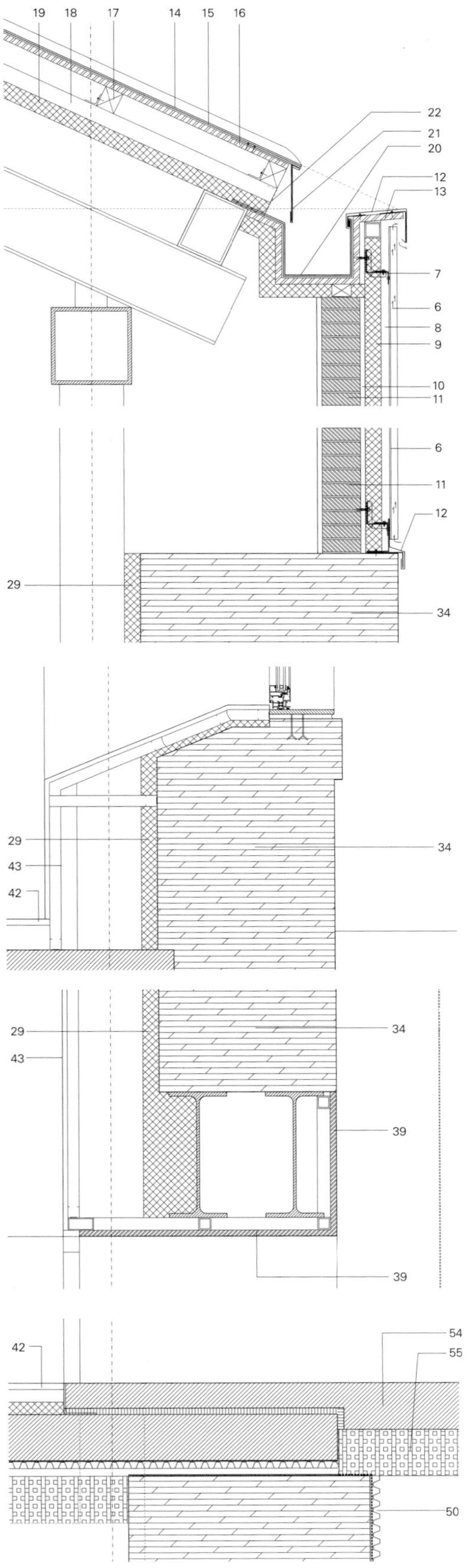

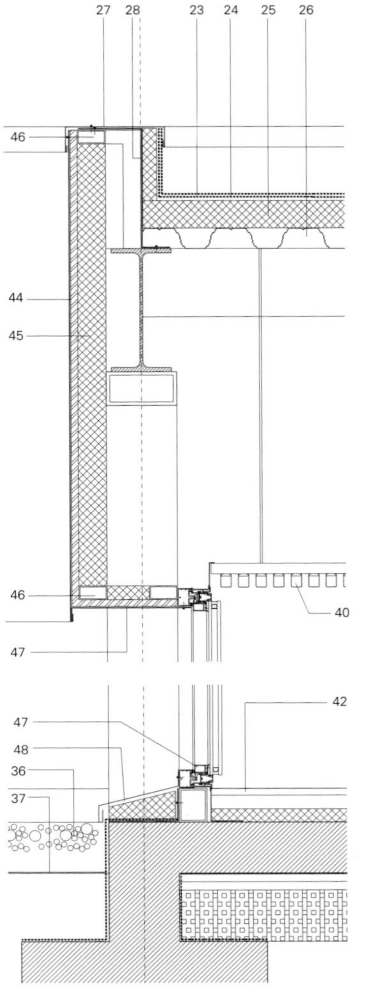

复活节雕塑博物馆

CALLE

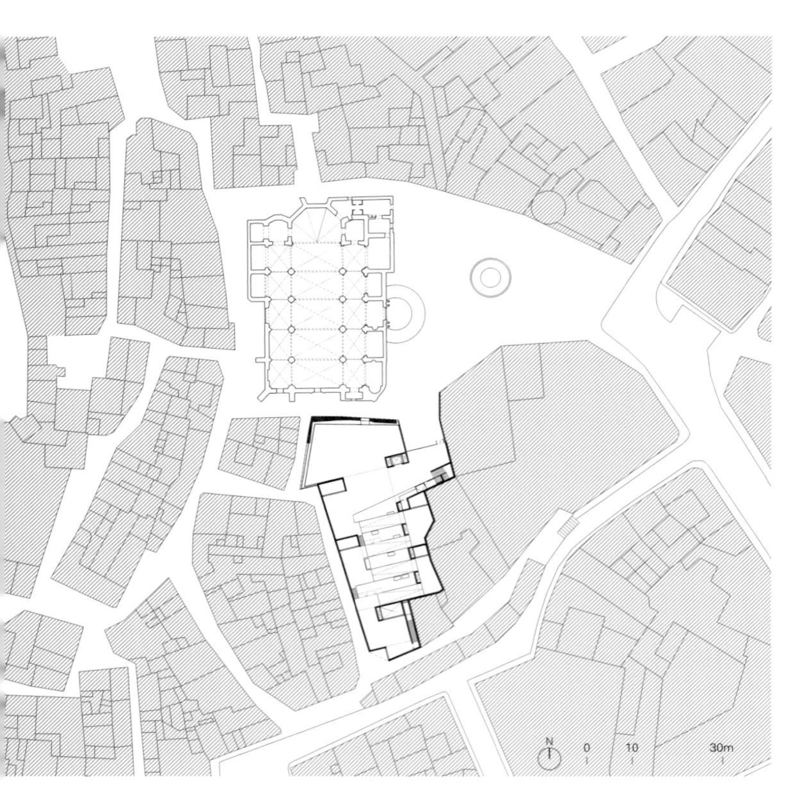

西班牙天主教徒的复活节庆祝活动要持续一周（圣周），其中包括濯足节和耶稣受难日游行。他们围绕"pasos"雕像（真人大小的木质雕像，描述耶稣走向十字架的过程及其他类似的行为）排成队伍，肩扛雕像，在拥挤的中世纪时代的市中心街道上前行。

为了恢复日益衰退的Casa del Conde（一座由佛罗里达布兰卡伯爵建造的18世纪不朽建筑）旁一个城市地块昔日的活力，Hellín小镇（西班牙卡斯蒂利亚–拉曼查，阿尔瓦塞特省，人口31 000）要求举办一场设计竞赛建造一座博物馆，希望收集宗教上的及有关复活节庆祝活动有纪念意义的东西，复活节庆祝活动也被称为"Tamboradas"（西班牙政府正式宣布其为国际旅游活动，并被列为国家遗产的一部分）。

最初Exit建筑师事务所提议把旧建筑恢复原貌，保留其庭院原来的样式，相应地占用一些新场地。但是，当得到委员会授权，经过一些额外的技术研究之后发现需要拆除旧建筑，只能保留其主立面的一小部分。

基地沿着Nuestra Señora del Rosario教堂（古摩尔时代的清真寺）周围的广场向外倾斜，获胜的同心状的方案转变成了一座巨大的建筑占据了整个地块。一层是一个单独的开放式空间，一系列长方形、不规则四边形木质家具沿着倾斜的地面陈列着，界定了之字形的行走路线。一层下面有一个地下室，上面有一个像夹层一样的二层，真正的博物馆上方还有一层屋顶，附加展区、储藏室、办公室及服务空间围绕着主大厅设置。

从形态学上看，主天井变成了更小的采光井，靠着邻近房屋墙壁中间，成为每个展览"岛"的开端和一条垂直街道末端的标志。

在合约当中最大的一部分是尽量把该项目和教堂广场连接起来，在旧建筑仅存的墙体后面设置一个中间大厅，博物馆入口就设在这里。

忏悔者们要扛着巨大的"pasos"雕像，慢慢行进在鹅卵石铺就的道路上，而游客们（以前游行道路上的观众）只需漫步在博物馆内如今固定好的雕像当中，光线通过建筑四周巨大的玻璃表面和混凝土盒式结构上切割出的洞口进入建筑，使建筑内部光线充足，此时博物馆显得很有诗意。

尽管从外部看起来该建筑显得很低调，但是它的表面和中世纪风格的相邻建筑仍然十分协调，甚至后墙外层还使用了石头，这样当人们走过这条街道的时候还会想起古教堂的墙壁。

像萨贡托的格拉西博物馆一样，Hellín的展厅不仅是对建筑类型的恢复，还给老城区增添了一种风格，在教堂里应用了只有在严格的建筑类型里才有的建筑形式，即使在风格迥异的相邻建筑间也会营造出一种谦逊的感觉，这让人感觉更舒适。

Easter Sculpture Museum

Easter celebrations among Spanish Catholics extend through a (Holy) week, and include Maundy Thursday and Good Friday processions, organized around the figure of "pasos" (life-size wooden statues, depicting Jesus' steps towards crucifixion and other similar motives), carried on the shoulders of brotherhood members through the packed streets of Medieval city centres.

Hoping to recover an urban plot adjacent to the decaying Casa del Conde (an 18th century monument built by the count of Floridablanca), the municipality of Hellín (Albacete province, Castilla – La Mancha, Spain, population 31,000) requested entries for a Museum, expecting to collect religious and commemorative objects allusive to Easter celebrations, also called Tamboradas (and officially declared by the Spanish government as events of International Touristic Interest, being part of the nation's heritage).

Exit Architects initially proposed to recover the old house, conserving its original courtyard typology and occupying the site accordingly. Nonetheless, once the commission was awarded, additional technical studies justified the total demolition of the building, from which only a fragment of the main facade was preserved.

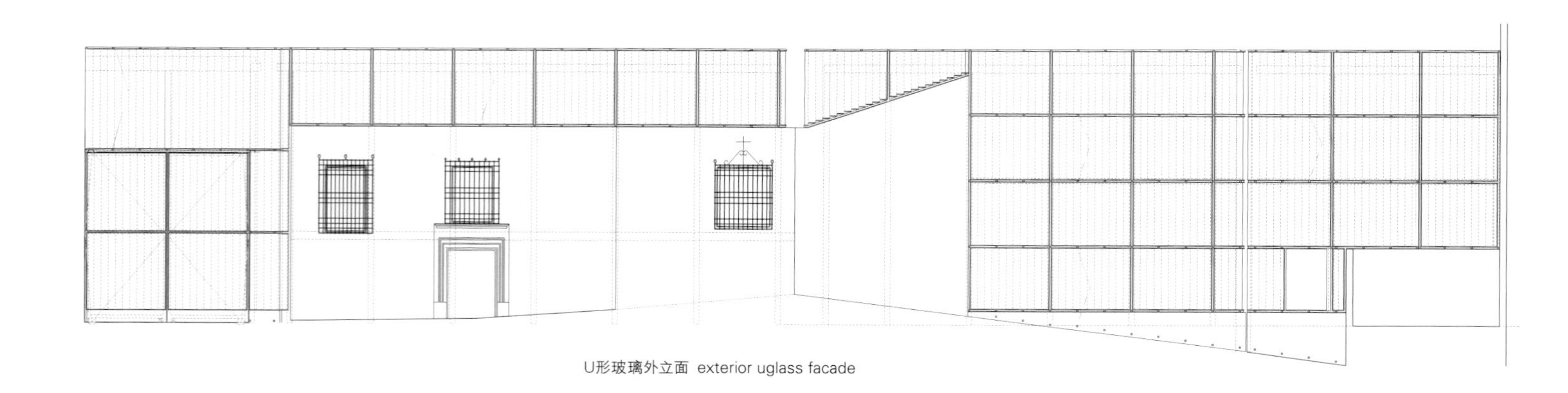

U形玻璃外立面 exterior uglass facade

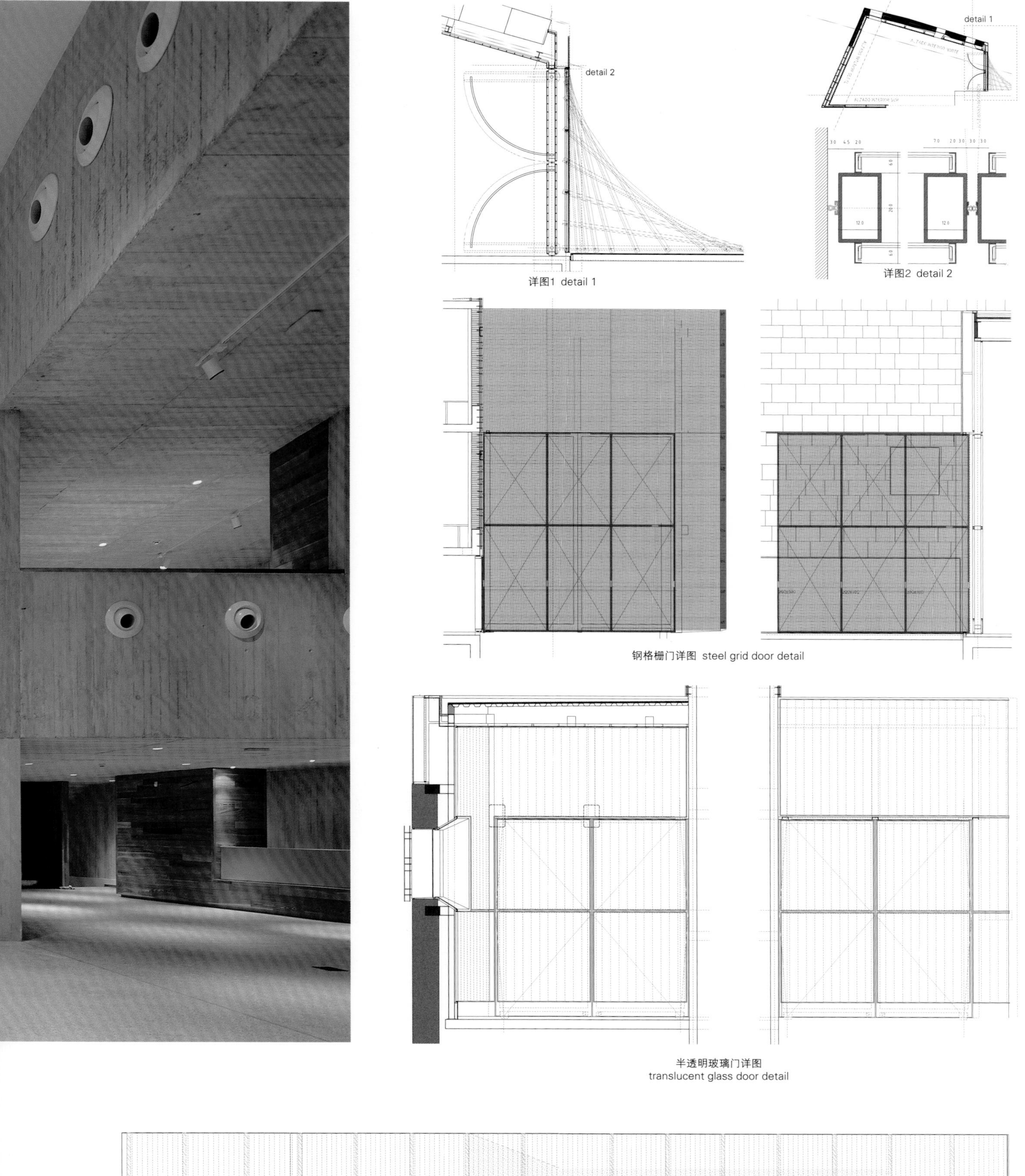

详图1 detail 1

详图2 detail 2

钢格栅门详图 steel grid door detail

半透明玻璃门详图
translucent glass door detail

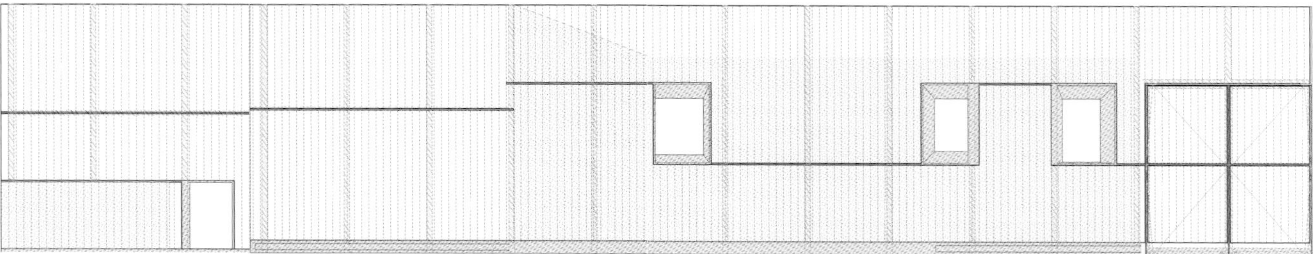

U形玻璃内立面 interior uglass facade

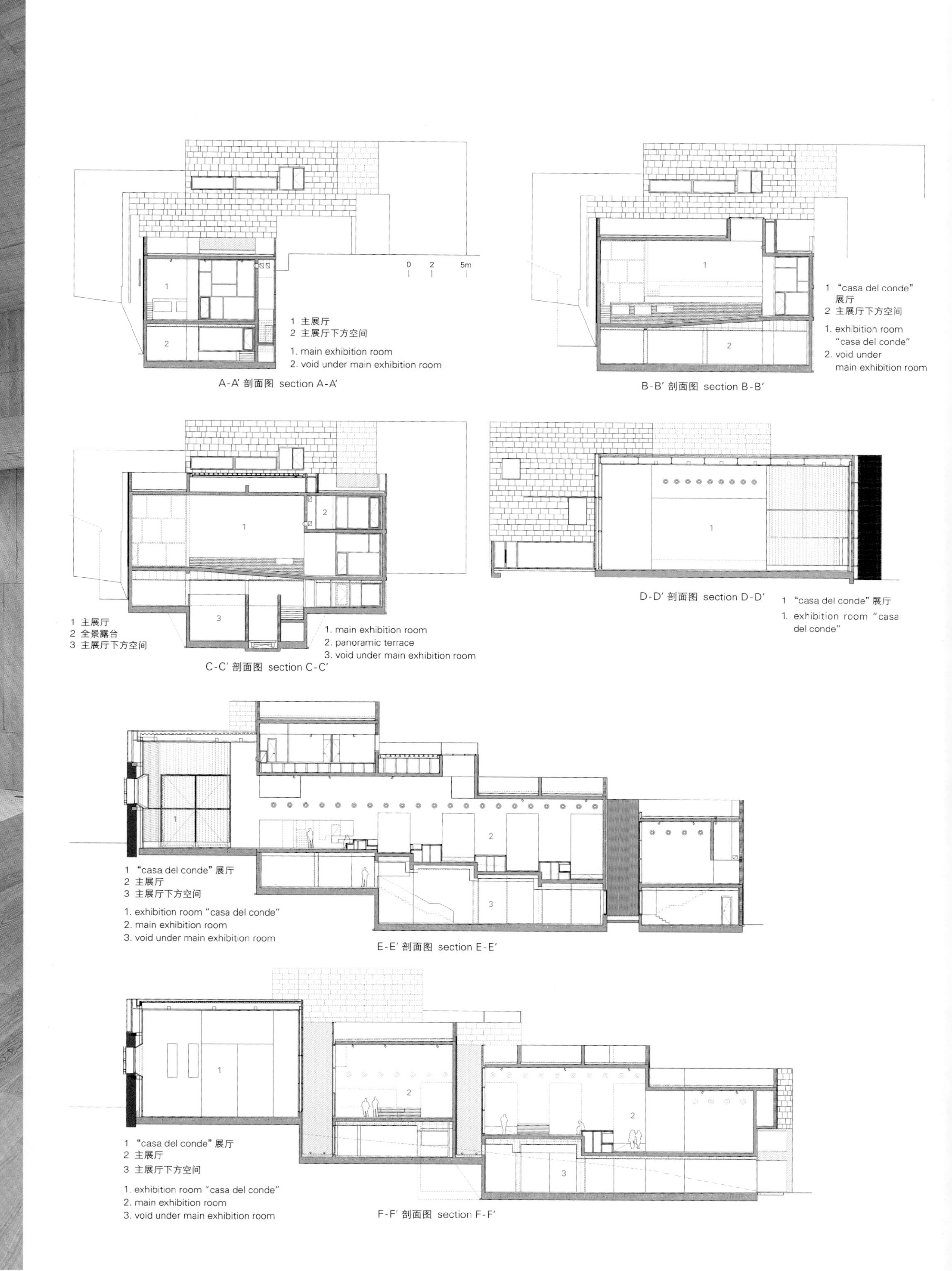
0 2 5m
1
2
1 主展厅
2 主展厅下方空间
1. main exhibition room
2. void under main exhibition room
A-A' 剖面图 section A-A'
1
2
1 "casa del conde" 展厅
2 主展厅下方空间
1. exhibition room "casa del conde"
2. void under main exhibition room
B-B' 剖面图 section B-B'
1
2
3
1 主展厅
2 全景露台
3 主展厅下方空间
1. main exhibition room
2. panoramic terrace
3. void under main exhibition room
C-C' 剖面图 section C-C'
1
D-D' 剖面图 section D-D'
1 "casa del conde" 展厅
1. exhibition room "casa del conde"
1
2
3
1 "casa del conde" 展厅
2 主展厅
3 主展厅下方空间
1. exhibition room "casa del conde"
2. main exhibition room
3. void under main exhibition room
E-E' 剖面图 section E-E'
1
2
3
1 "casa del conde" 展厅
2 主展厅
3 主展厅下方空间
1. exhibition room "casa del conde"
2. main exhibition room
3. void under main exhibition room
F-F' 剖面图 section F-F'

1 临时展厅
2 行政部门
3 全景露台

1. temporary exhibition room
2. administration
3. panoramic terrace

二层 second floor

1 入口
2 "casa del conde" 展厅
3 主展厅

1. entrance
2. exhibition room "casa del conde"
3. main exhibition room

一层 first floor

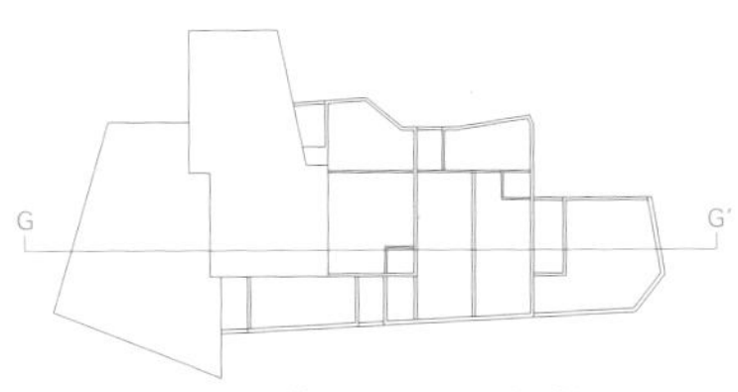

G-G' 剖面图 section G-G'

1. gutter: double galvanized steel sheet thick=1.5 mm. isolating polyurethane foam. fixation of exterior sheet to facade substructure (overlap 15 cm).
2. roof exterior: galvanized steel sheet thick=1.5 mm, piece 1: fixed to facade substructure allowing ventilation, piece 2: fixed to roof steel sheet
3. roof interior: lacquered steel sheet thick=1.5 mm. fixed to substructure
4. upper part existing wall: galvanized steel sheet thick=1.5 mm. encounter fixed with sikaflex 11 fc or similar.
5. loler part existing wall: brick wall f3, plasterdon its outer side. on it leans the substructure of facade ug2.
6. upper part stone facade: galvanized steel sheet thick=1.5 mm. mechanical fixation to brick wall.
7. non transit roof: polyethylene e=2 cm between concrete walls and roof horizontal overlap 30 cm. vertical overlap 15 cm upon gravel level
8. air conditioning in-flow: nozzles according to engineering plans. holes according to structure plans
9. uglass perimeter: anodized aluminum uf 60.20.2 dull silver. fixed substructure. sealed with sikaflex 11 fc or similar.
10. facade substructure: galvanized steel tube #120.60.3/ 60.60.6. welded to heb200.
11. stone roof: support with brick walls 5 cm perimeter hole for ventilation, covered with steel sheet to avoid rain water income
12. handrail: metallic profile ld200.100.10 fixed on the vertical side of the concrete/brick wall.
13. cabinets: air conditioning return in the place of the skirting
14. concrete beams: in the axis n° 3 the beam will be 5 cm set backwards, so that the wood is in the same vertical line as the concrete pillar
15. steel sheet facade: substructure steel tube #40.3. isolating polyurethane foam e=5 cm. galvanized steel sheet ansi 304/2b "small wave".
16. roof: above the concrete roof it will be again used the grey concrete for the beams
17. window : galvanized steel tube, mechanical fixed to the concrete ceiling will be 10 cm set backwards to allow the galvanized steel sheet to pass by
18. drain material: epdm. protection for the gravel reinforced waterproof sheets
19. on roof slab of exhibition floors: apartir of the upper bound of the slabs that form the roof of the main room (+6.80/+4.80/+2.80) is executed supporting elements not seen with Salco HG1 concrete type wall lift, the staircase and the skylights. in any case, the element somprobara concreting with HG1 not stay cysto asking the checking by the df
20. 11trasdosado: thermal insulation in areas of administration according to plan c05 details. insulation board is placed inside rigid price larealizacion the cladding.

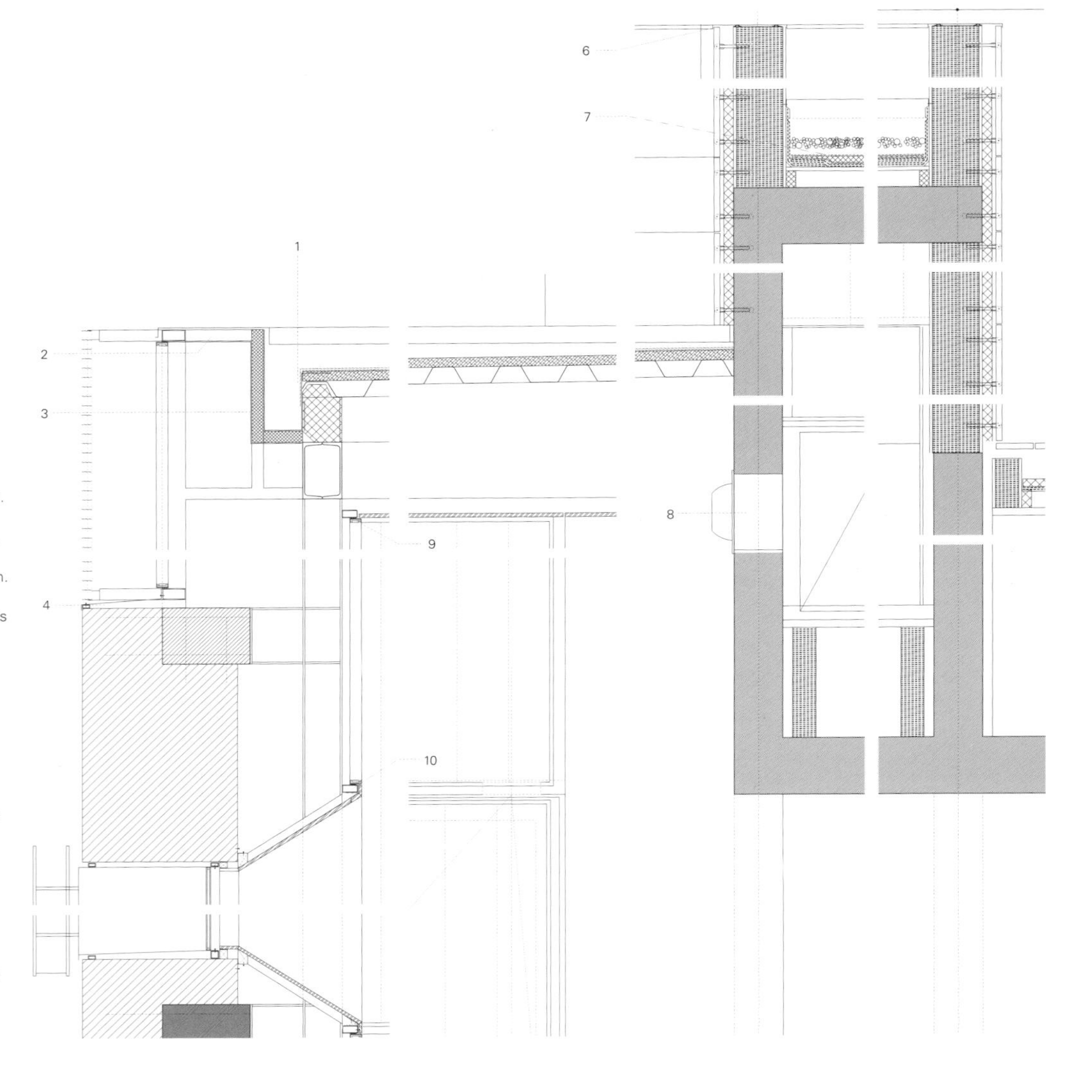

G-G' 剖面详图
section G-G' detail

Sloping away from the square formed around the Nuestra Señora del Rosario Church (built over the old Moorish mosque), the winning concentric scheme turned into a massive block, occupying the full extent of the plot. The resulting space on the ground floor is a single open plan, in which wooden furniture – a set of elongated, trapezoidal cupboards – marks a zigzag route, ramping down with the site's inclination. Above and below, partially occupying a basement and a mezzanine-like first floor, and creating a roof over the actual museum, additional exhibition, storage, office, and service space sandwich the main hall.

At a morphological level, a main patio has turned into smaller light wells, located against the median walls of neighboring houses, marking the beginning of each exhibition "island" or the end of a perpendicular street. The largest of all these indentures, though, tries to tie the project to the Church's plaza, creating an intermediate hall that marks the entrance to the Museum, behind the only remaining wall of the old house.

While fraternity penitents slowly walk along the cobbled stones, bearing the weight of the massive "pasos" on their shoulders, this project proposes a rather poetic inversion of the scheme, by making visitors (former spectators of the procession's path) stroll themselves among the now fixed iconography, in a space deliberately filled with light by enormous glazed surfaces around the blocks' perimeter, and through holes carved into the concrete box.

On the outside, though, the building remains rather silent, keeping its surfaces as tense as those of its medieval neighbors, and even camouflaging behind stone cladding reminiscent of the walls of the old church across the street.

More than a typological restitution, such as that proposed by Grassi at Sagunto, the exhibition hall at Hellín becomes a stylistic addition to the old town, inserting an architectural form only comparable in strict typological terms to the church in front of it, while creating the illusion of modesty that allows it to feel comfortable, even among very different neighbors. Jorge Alberto Mejía Hernández

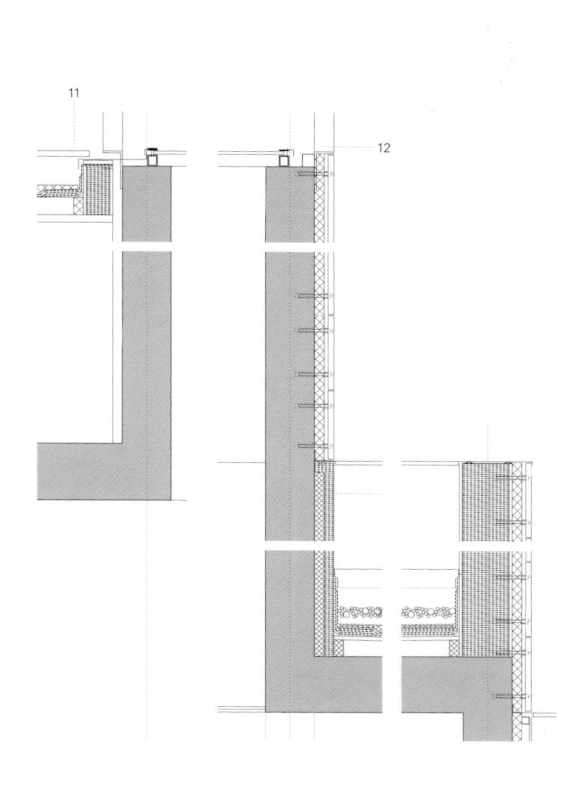

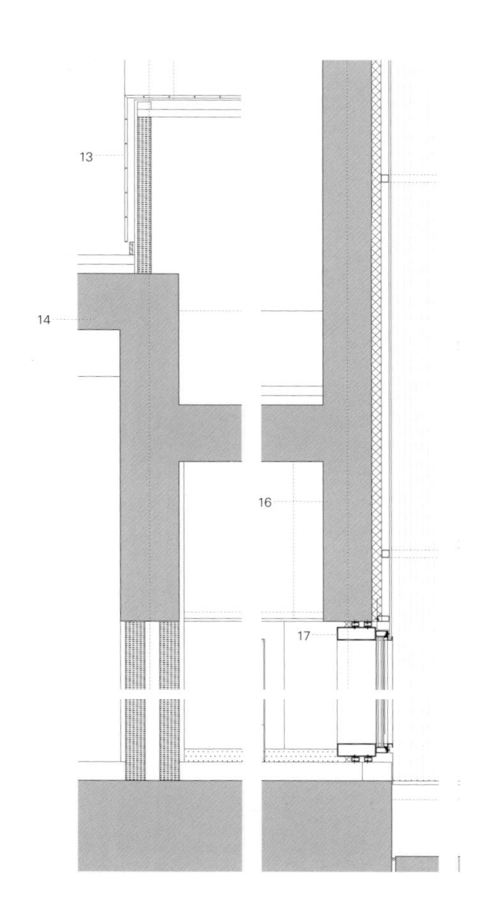

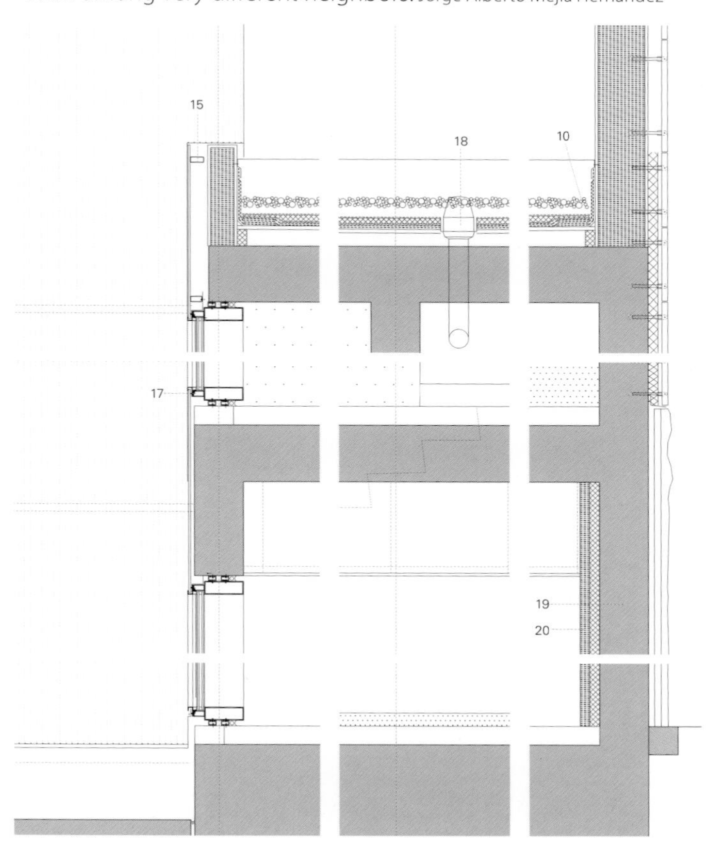

项目名称：Easter Sculpture Museum, Hellín, Albacete
地点：Hellín, Albacete, Spain
建筑师：Ibán Carpintero, Mario Sanjuán
合作者：Miguel García-Redondo, Silvia N. Gómez, Ángel Sevillano, José Mª Tabuyo
技术建筑师：Alberto Palencia, José Antonio Alonso
结构顾问：INDAGSA (José Luis Cano)
机械顾问：Maintenance Ibérica
总承包商：PEFERSAN, S.A.
甲方：Public Works Ministry / Hellín Municipality
用地面积：1236.95m² 建筑面积：2,160m² 总楼面面积：950.35m²
造价：EUR 3,512,235
设计时间：2002 竣工时间：2011
摄影师：©FG+SG Architectural Photography

>>26

Circa Morris-Nunn Architects

Robert Morris Nunn is a director and principal architect of Circa Morris-Nunn Architects. Has practiced in Tasmania for over 25 years, and has taken a special interest in the social impact of architecture and collaborative design process. The basic concept of his work has been to explore new approaches towards a more enlightened, socially humane architecture which as part of their resolution have often integrated innovative structural ideas. Has received great international interest – most recently in the field of ecologically sustainable design. Has won 3 national architectural awards and over 40 awards, his work has been illustrated in international publications. Has been invited to lecture at many Australian and international conferences about his work. Was recently appointed as adjunct professor at the School of Architecture, University of Tasmania.

>>136

Roberto Puchetti

Obtained his diploma in 1995 at the Central University of Venezuela. At the same university in 2004, obtained a Master of Science on Architectonic Design. More recently he was awarded Best Venezuelan Project at BIAU 2010 – 7th Iberoamericano Biennale of Architecture and Urbanism, celebrated in Medellín (Colombia) for the New Vaccines Production Laboratory of the National Institute of Hygiene, Rafael Rangel. Taught at the Central University of Venezuela. Has also been invited as a teacher at the Pontifical Catholic University of Chile. Currently teaches at the architectural design studio at the Simon Bolivar University.

>>144

Dekleva Gregoric Arhitekti

Was founded in 2003 by Aljosa Dekleva[left] and Tina Gregoric[right] in Ljubljana, Slovenia. Both of them graduated from University of Ljubljana, Slovenia and continued their study at the Architectural Association School of Architecture (AA school) where they received Master Degree in Architecture with Distinction in 2002. Have been visiting lecturers and critics at the AA School, Technical University of Graz Austria and many others. Understand design as research on several modes (social, material, and historical) and respond to specific constrains and conditions. Rather than a conceptual approach to intense structuring of space, they challenge the use of materials and the exposure of their natures. Are aiming to stimulate new social interactions among users, participation of the users', design process and customization to users' needs.

>>72

Satoshi Okada Architects

Satoshi Okada was born in Japan in 1962. After he graduated from GSAPP Columbia University, received Ph.D. at Waseda University in 1993. Established Satoshi Okada Architects in 1995 in Tokyo. Currently teaches Design and Theory at Graduate School of Architecture, Chiba University in Japan. Since 2010, has been a visiting professor at the University of Venice (IUAV) and a research fellow at the Toyota Foundation. His works have been published in numerous books and magazines around the world.

>>128

Atelier Carvalho Araujo

Was founded in 1996 by José Manuel Castro Carvalho Araújo. He was born in Braga, Portugal in 1961. Took over the management of the Carvalho Araújo Study Centre in 1986 which functions as a design laboratory, testing a new ways of looking at industrial design issues. In 1990, he completed his degree in architecture at the School of Architecture Technical University of Lisbon(FAUTL). Since 2005, he has been lecturing with his practice, teaching Project Design in the 5th year of the Degree at the School of Architecture University of Catolic.

Frequently invited to events related to project design, conferences, juries and exhibitions. Received major International Awards for the excellence in the performance of his professional activity.

>>34

Herault Arnod Architectes

Was founded by Yves Arnod[right] and Isabel Hérault[left] in 1991. They work on a various programs on different scales, refuting any idea of specialization. Their goal is to escape from the ready-made solutions or the transposition of receipts. It leads them to original answers which draw their substance from the reality of situation. Have been practicing in the cultural fields including theatres, concert halls, recording studios, rehearsal rooms, media libraries and museums. Has developed a working method for the strict control of the costs, schedules and the quality of the realization. Believes that they exert their creativity and architectural research in this framework.

>>102

X Architekten

Was established in 1996 in Graz. Has been managed by Max Nirnberger, Bettina Brunner, Lorenz Prommegger, David Birgmann and Rainer Kasik. Is a group of committed architects who develop conceptual positions on contemporary architecture through project-related work. As a mathematical variable, X stands for openness. X demands the plural: the team with its flat hierarchy replaces the professional profile of the architect as lone fighter. The dynamics of a permanent work process oscillating between creativity and (self-) criticism permit the emergence of a quality standard above the capacity of each single member.

The activities of X architekten comprise the full range from housing and office construction, buildings for industry and commerce, shopping and entertainment projects to design and urbanism.

>>82

ALA Architects

Is a Helsinki based architecture firm which has been in practice by Samuli Woolston, Antti Nousjoki, Janne Teräsvirta and Juho Grönholm since 2004. They try to seek fresh angles, flowing forms and new solutions on all levels of architecture. Has focused on developing alternative prototypes. Believes in creating beauty with a design approach which combines the intuitive, analytic and the practical with the extravagant.

Currently works on projects ranging from underground stations to housing areas, high-rise developments and culture venues.

Pereda Pérez Arquitectos

Was founded by Carlos Pereda Iglesias and Oscar Perez Silanes in Pamplona, where they conduct their business engaged in the research and realization of architectural projects.

Their works has been published in various journals and national publications. Are developing their own teaching methods in ETS UNAV(1996-2001), ESCARQ, the UIC in Barcelona, the University of Alicante and Las Palmas de Gran Canaria and others.

Atelier d'architectes Fournier-Maccagnan

Pascal Fournier[left] was born in 1972. Studied architecture and graduated from the university of Freiburg in 1997. Founded his own studio in 1999 which became Atelier d'architectes Fournier-Maccagnan in 2000.

Sandra Maccagnan[right] was born in 1975. Graduated from the University of Freiburg in 1997 and joined Atelier d'architectes Fournier-Maccagnan in 2000.

>>112

Flexo Arquitectura

Was set up in Barcelona in 2002 by Tomeu Ramis Frontera[left], Aixa del Rey García[middle] and Bárbara Vich Arrom[right]. They were all born in Palma de Mallorca. Graduated from the ETSA in Barcelona(ETSAB). Has been prizewinners in many competitions and their works have been published in numerous publications. Has given lectures in dfferent institutions and the office has been awarded in several occasions. Tomeu Ramis currently teaches at the Swiss Federal Institute of Technlogy Zurich(ETH Zürich) as assistant professor.

Aldo Vanini
Practices in the fields of Architecture and Planning. Had many of his works published in various qualified international magazines. Is a member of regional and local government boards, involved in architectural and planning researches. One of his most important research interests is the conversion of abandoned mining sites in Sardinia.

Alison Killing
Is an architect and urban designer based in Rotterdam, the Netherlands. Has written for several architecture and design magazines in the UK, contributing features and reviews to *Blueprint* and *Icon* and editing the research section of *Middle East Art Design and Architecture*. Most recently, she has worked as a correspondent for the online sustainability magazine *Worldchanging*. Alison has an eclectic design background, ranging from complex geometry and structural engineering, to humanitarian practice, to architecture and urban design and has worked internationally, in the UK and the Netherlands, but also more widely in Europe, Switzerland, China and Russia.

Jorge Alberto Mejia Hernandez
Had his education as an Architect at the Universidad del Valle and graduated in 1996. Holds a Master in History and Theory of Art and Architecture (2002) as well as a Master Degree in Architecture (2008), both from the Universidad Nacional de Colombia. His teaching includes Architectural Theory, History of Architecture and Design Studio at the Universidad Nacional de Colombia since February 2005, where he became Professor Catedratico Asociado in 2007.
Has written many books such as of Enrique Triana: Obras y Proyectos (Bogotá: Planeta, 2006) and Coauthor of Vivienda Moderna en Colombia (Bogotá: Universidad Nacional de Colombia, 2004) and XX Bienal Colombiana de Arquitectura (Bogotá: Sociedad Colombiana de Arquitectos, 2006). His research interests include architectural form, modern architecture, contemporary conditions and architectural principles and procedures.

>>62
Reiulf Ramstad Architects
Reiulf Ramstad was born in 1962 in Oslo, Norway. Studied at the University of Genoa and received Ph.D. from the University of Venice(IUAV). Founded his own office in 1995. Participated in more than 60 restricted and open competitions between 1993 and 2011. Has been awarded many prizes in Norway and abroad. His main activities related to education and research at Norwegian School of Science and Technology(NTNU) as assistant professor and research assistant.

>>154
Exit Architects
Was founded by Angel Sevillano and José Maria Tabuyo. Is staffed by a team of architects, including Iban Carpintero, Mario Sanjuan and Miguel Garcia Redondo from different parts of Spain who work together to integrate their knowledge and experience. Believes it is possible to integrate technical rationality and the emotion in space, to combine the construction precision and the intuition, and to unite functionality and beauty. Pursuits the optimum integration of all the circumstances of each project, the relation between the buildings and the city, the sustainability and the consequences of producing contemporary architecture.

图书在版编目(CIP)数据

建筑入景 : 汉英对照 / C3出版公社编 ； 赵姗姗等译.
— 大连 : 大连理工大学出版社, 2012.9
(C3建筑立场系列丛书; 19)
ISBN 978-7-5611-7306-0

Ⅰ. ①建… Ⅱ. ①C… ②赵… Ⅲ. ①建筑艺术一研究
一汉、英 Ⅳ. ①TU-80

中国版本图书馆CIP数据核字（2012）第219990号

出版发行：大连理工大学出版社
（地址：大连市软件园路 80 号 邮编：116023）
印 刷：精一印刷（深圳）有限公司
幅面尺寸：225mm×300mm
印 张：12
出版时间：2012 年 9 月第 1 版
印刷时间：2012 年 9 月第 1 次印刷
出 版 人：金英伟
统 筹：房 磊
责任编辑：杨 丹
封面设计：王志峰
责任校对：张媛媛

书 号：ISBN 978-7-5611-7306-0
定 价：228.00 元

发 行：0411-84708842
传 真：0411-84701466
E-mail: a_detail@dutp.cn
URL: http://www.dutp.cn